Communications
in Computer and Information Science   97

Kang Li   Xin Li   Shiwei Ma
George W. Irwin (Eds.)

Life System Modeling and Intelligent Computing

International Conference on Life System Modeling
and Simulation, LSMS 2010
and International Conference on Intelligent Computing
for Sustainable Energy and Environment, ICSEE 2010
Wuxi, China, September 17-20, 2010
Proceedings, Part I

 Springer

Volume Editors

Kang Li
Queen's University Belfast
Belfast, UK
E-mail: k.li@qub.ac.uk

Xin Li
Shanghai University, China
E-mail: su_xinli@yahoo.com.cn

Shiwei Ma
Shanghai University, China
E-mail: swma@mail.shu.edu.cn

George W. Irwin
Queen's University Belfast
Belfast, UK
E-mail: g.irwin@ee.qub.ac.uk

Library of Congress Control Number: 2010934237

CR Subject Classification (1998): J.3, H.2.8, F.1, I.6.1, I.6, I.2.9

ISSN   1865-0929
ISBN-10   3-642-15852-8 Springer Berlin Heidelberg New York
ISBN-13   978-3-642-15852-0 Springer Berlin Heidelberg New York

springer.com

© Springer-Verlag Berlin Heidelberg 2010

Typesetting: Camera-ready by author, data conversion by Scientific Publishing Services, Chennai, India
Printed on acid-free paper   06/3180   5 4 3 2 1 0

Preface

The 2010 International Conference on Life System Modeling and Simulation (LSMS 2010) and the 2010 International Conference on Intelligent Computing for Sustainable Energy and Environment (ICSEE 2010) were formed to bring together researchers and practitioners in the fields of life system modeling/simulation and intelligent computing applied to worldwide sustainable energy and environmental applications.

A life system is a broad concept, covering both micro and macro components ranging from cells, tissues and organs across to organisms and ecological niches. To comprehend and predict the complex behavior of even a simple life system can be extremely difficult using conventional approaches. To meet this challenge, a variety of new theories and methodologies have emerged in recent years on life system modeling and simulation. Along with improved understanding of the behavior of biological systems, novel intelligent computing paradigms and techniques have emerged to handle complicated real-world problems and applications. In particular, intelligent computing approaches have been valuable in the design and development of systems and facilities for achieving sustainable energy and a sustainable environment, the two most challenging issues currently facing humanity. The two LSMS 2010 and ICSEE 2010 conferences served as an important platform for synergizing these two research streams.

The LSMS 2010 and ICSEE 2010 conferences, held in Wuxi, China, during September 17–20, 2010, built upon the success of two previous LSMS conferences held in Shanghai in 2004 and 2007 and were based on the Research Councils UK (RCUK)-funded Sustainable Energy and Built Environment Science Bridge project. The conferences were jointly organized by Shanghai University, Queen's University Belfast, Jiangnan University and the System Modeling and Simulation Technical Committee of CASS, together with the Embedded Instrument and System Technical Committee of China Instrument and Control Society. The conference program covered keynote addresses, special sessions, themed workshops and poster presentations, in addition to a series of social functions to enable networking and foster future research collaboration.

LSMS 2010 and ICSEE 2010 received over 880 paper submissions from 22 countries. These papers went through a rigorous peer-review procedure, including both pre-review and formal refereeing. Based on the review reports, the Program Committee finally selected 260 papers for presentation at the conference, from amongst which 66 were subsequently selected and recommended for publication by Springer in two volumes of *Communications in Computer and Information Science* (CCIS). This particular volume of *Communications in Computer and Information Science* (CCIS) includes 32 papers covering 7 relevant topics.

The organizers of LSMS 2010 and ICSEE 2010 would like to acknowledge the enormous contributions from the following: the Advisory and Steering Committees for their guidance and advice, the Program Committee and the numerous referees worldwide for their significant efforts in both reviewing and soliciting the papers, and the Publication Committee for their editorial work. We would also like to thank

Alfred Hofmann, of Springer, for his continual support and guidance to ensure the high-quality publication of the conference proceedings. Particular thanks are of course due to all the authors, as without their excellent submissions and presentations, the two conferences would not have occurred.

Finally, we would like to express our gratitude to the following organizations: Chinese Association for System Simulation (CASS), IEEE SMCS Systems Biology Technical Committee, National Natural Science Foundation of China, Research Councils UK, IEEE CC Ireland chapter, IEEE SMC Ireland chapter, Shanghai Association for System Simulation, Shanghai Instrument and Control Society and Shanghai Association of Automation.

The support of the Intelligent Systems and Control research cluster at Queen's University Belfast, Tsinghua University, Peking University, Zhejiang University, Shanghai Jiaotong University, Fudan University, Delft University of Technology, University of Electronic Science Technology of China, Donghua University is also acknowledged.

July 2010

Bohu Li
Mitsuo Umezu
George W. Irwin
Minrui Fei
Kang Li
Luonan Chen
Xin Li
Shiwei Ma

LSMS-ICSEE 2010 Organization

Advisory Committee

Kazuyuki Aihara, Japan
Zongji Chen, China
Guo-sen He, China
Frank L. Lewis, USA
Marios M. Polycarpou,
   Cyprus
Olaf Wolkenhauer,
   Germany
Minlian Zhang, China

Shun-ichi Amari, Japan
Peter Fleming, UK
Huosheng Hu,UK
Stephen K.L. Lo, UK

Zhaohan Sheng, China

Cheng Wu, China
Guoping Zhao, China

Erwei Bai, USA
Sam Shuzhi Ge, Singapore
Tong Heng Lee, Singapore
Okyay Kaynak, Turkey
Peter Wieringa,
   The Netherlands

Yugeng Xi, China

Steering Committee

Sheng Chen, UK
Tom Heskes,
   The Netherlands
Zengrong Liu, China
MuDer Jeng, Taiwan, China
Kay Chen Tan, Singapore
Haifeng Wang, UK
Guangzhou Zhao, China

Kwang-Hyun Cho, Korea

Shaoyuan Li, China
Sean McLoone, Ireland
Xiaoyi Jiang, Germany
Kok Kiong Tan, Singapore
Tianyuan Xiao, China
Donghua Zhou, China

Xiaoguang Gao, China

Liang Liang, China
Robert Harrison, UK
Da Ruan, Belgium
Stephen Thompson, UK
Jianxin Xu, Singapore
Quanmin Zhu, UK

Honorary Chairs

Bohu Li, China
Mitsuo Umezu, Japan

General Chairs

George W. Irwin, UK
Minrui Fei, China

International Program Committee

IPC Chairs

Kang Li, UK
Luonan Chen, Japan

IPC Regional Chairs

Haibo He, USA
Wen Yu, Mexico
Shiji Song, China
Xingsheng Gu, China
Ming Chen, China

Amir Hussain, UK
John Morrow, UK
Taicheng Yang, UK
Yongsheng Ding, China
Feng Ding, China

Guangbin Huang, Singapore
Qiguo Rong, China
Jun Zhang, USA
Zhijian Song, China
Weidong Chen, China

IPC Members

Maysam F. Abbod, UK
Vitoantonio Bevilacqua,
    Italy
Yuehui Chen, China

Minsen Chiu, Singapore
Kevin Curran, UK
Jianbo Fan, China

Huijun Gao, China
Xudong Guo, China
Haibo He, USA
Fan Hong, Singapore
Yuexian Hou, China
Guangbin Huang,
    Singapore
MuDer Jeng, Taiwan,
    China

Yasuki Kansha, Japan
Gang Li, UK
Yingjie Li, China
Hongbo Liu, China
Zhi Liu, China
Fenglou Mao, USA
John Morrow, UK
Donglian Qi, China
Chenxi Shao, China
Haiying Wang, UK
Kundong Wang, China
Wenxing Wang, China
Zhengxin Weng, China
WeiQi Yan, UK
Wen Yu, Mexico
Peng Zan, China
Degan Zhang, China
Huiru Zheng, UK
Huiyu Zhou, UK

Peter Andras, UK
Uday K. Chakraborty,
    USA
Xinglin Chen, China

Michal Choras, Poland
Mingcong Deng, Japan
Haiping Fang, China
Wai-Keung Fung, Canada
Xiao-Zhi Gao, Finland
Aili Han, China
Pheng-Ann Heng, China
Xia Hong, UK
Jiankun Hu, Australia

Peter Hung, Ireland

Xiaoyi Jiang, Germany
Tetsuya J. Kobayashi,
    Japan
Xiaoou Li, Mexico
Paolo Lino, Italy
Hua Liu, China
Sean McLoone, Ireland
Kezhi Mao, Singapore
Wasif Naeem, UK
Feng Qiao, China
Jiafu Tang, China
Hongwei Wang, China
Ruisheng Wang, USA
Yong Wang, Japan
Lisheng Wei, China
Rongguo Yan, China
Zhang Yuwen, USA
Guofu Zhai, China
Qing Zhao, Canada
Liangpei Zhang, China
Shangming Zhou, UK

Costin Badica, Romania

Tianlu Chen, China
Weidong Cheng, China
Tommy Chow,
    Hong Kong, China
Frank Emmert-Streib, UK
Jiali Feng, China
Houlei Gao, China
Lingzhong Guo, UK
Minghu Ha, China
Laurent Heutte, France
Wei-Chiang Hong, China
Xiangpei Hu, China

Amir Hussain, UK

Pingping Jiang, China

Aim`e Lay-Ekuakillel, Italy
Xuelong Li, UK
Tim Littler, UK
Wanquan Liu, Australia
Marion McAfee, UK
Guido Maione, Italy
Mark Price, UK
Alexander Rotshtein, Ukraine
David Wang, Singapore
Hui Wang, UK
Shujuan Wang, China
Zhuping Wang, China
Ting Wu, China
Lianzhi Yu, China
Hong Yue, UK
An Zhang, China
Lindu Zhao, China
Qingchang Zhong, UK

Secretary-General

Xin Sun, China
Ping Zhang, China
Huizhong Yang, China

Publication Chairs

Xin Li, China
Wasif Naeem, UK

Special Session Chairs

Xia Hong, UK
Li Jia, China

Organizing Committee

OC Chairs

Shiwei Ma, China
Yunjie Wu, China
Fei Liu, China

OC Members

Min Zheng, China	Banghua Yang, China	Yang Song, China
Yijuan Di, China	Weihua Deng, China	Tim Littler, UK
Qun Niu, UK	Xianxia Zhang, China	

Reviewers

Renbo Xia, Vittorio Cristini, Aim'e Lay-Ekuakille, AlRashidi M.R., Aolei Yang, B. Yang, Bailing Zhang, Bao Nguyen, Ben Niu, Branko Samarzija, C. Elliott, Chamil Abeykoon, Changjun Xie, Chaohui Wang, Chuisheng Zeng, Chunhe Song, Da Lu, Dan Lv, Daniel Lai, David Greiner, David Wang, Deng Li, Dengyun Chen, Devedzic Goran, Dong Chen, Dongqing Feng, Du K.-L., Erno Lindfors, Fan Hong, Fang Peng, Fenglou Mao, Frank Emmert-Streib, Fuqiang Lu, Gang Li, Gopalacharyulu Peddinti, Gopura R. C., Guidi Yang, Guidong Liu, Haibo He, Haiping Fang, Hesheng Wang, Hideyuki Koshigoe, Hongbo Liu, Hongbo Ren, Hongde Liu, Hongtao Wang, Hongwei Wang, Hongxin Cao, Hua Han, Huan Shen, Hueder Paulo de Oliveira, Hui Wang, Huiyu Zhou, H.Y. Wang, Issarachai Ngamroo, Jason Kennedy, Jiafu Tang, Jianghua Zheng, Jianhon Dou, Jianwu Dang, Jichun Liu, Jie Xing, Jike Ge, Jing Deng, Jingchuan Wang, Jingtao Lei, Jiuying Deng, Jizhong Liu, Jones K.O., Jun Cao, Junfeng Chen, K. Revett, Kaliviotis Efstathios, C.H. Ko, Kundong Wang, Lei Kang,

Table of Contents – Part I

The First Section: Intelligent Modeling, Monitoring, and Control of Complex Nonlinear Systems

The Second Section: Modeling and Simulation of Societies and Collective Behaviour

The Third Section: Advanced Theory and Methodology in Fuzzy Systems and Soft Computing

The Fourth Section: Biomedical Signal Processing, Imaging, and Visualization

The Fifth Section: Computational Intelligence in Utilization of Clean and Renewable Energy Resources

The Sixth Section: Innovative Education for Sustainable Energy and Environment

The Seventh Section: Intelligent Methods in Power and Energy Infrastructure Development

Table of Contents – Part II

The Second Section: Advanced Neural Network Theory and Algorithms

The Third Section: Innovative Education in Systems Modeling and Simulation

The Fourth Section: Intelligent Methods in Developing Vehicles, Engines and Equipments

The Fifth Section: Fuzzy, Neural, and Fuzzy-Neuro Hybrids

Research on Steam Generator Water Level Control System Based on Nuclear Power Plant Simulator

Jianghua Guo

School of Power and Mechanical Engineering, Wuhan University
Wuhan, Hubei, China 430072
haihua2@163.com

Abstract. Steam generator (SG) is one of the most important equipments in nuclear power plants. The water level of SG must be kept in a certain range to ensure the plants operate safely, reliably and economically. Nowadays, most SG water levels are controlled by PID in PWR plants. In this paper, the mathematical models of SG level control system are built by Matlab/Simulink; the simulation research based on the Matlab/Simulink models is conducted, too. Then, based on RINSIM simulation platform of nuclear power plant Simulator in Wuhan University, a simulation model on the SG level control system is established. The graphical modeling methods for the SG level control system are provided. By using the model, transient simulation experiments and researches with different conditions are conducted. Contrast with the Matlab/Simulink simulation models, the good preciseness and identification performance of the RINSIM models are verified.

Keywords: Nuclear power plant Simulator, Steam generator, Level control system, Transient simulation.

1 Introduction

Steam generator is a heat interchanger equipment to generate steam which the turbine needs .In a nuclear reactor, energy generated by the nuclear fission is taken away by the coolant and transferred to the feed water of the secondary system through the steam generator to maintain the steam a certain pressure, temperature and dryness. The flue goes into the steam turbine to do work which is then converted into electrical or mechanical energy. In this energy conversion process, the steam generator is a part of the primary loop and the second loop in the same time, as it is called the pivot of the primary and the second circuit.

In the nuclear plant, the water level of steam generator is an important operating parameter. If the water level is too low, as a result, the steam will enter the water ring, which has a risk to cause the water hammer. While the water level is too high, it will drown the separator or even the dryer. In addition, if the humidity of the export saturated vapor is too high, it will accelerate the erosion of the steam turbine blade. Besides, as the controlled object has a system slowdown, a false water level will arise whereas there are changes in operating conditions.

K. Li et al. (Eds.): LSMS / ICSEE 2010, Part I, CCIS 97, pp. 1–7, 2010.

When there is a positive step change in steam flow, the water level of the steam generator will rise up in the early time, after reaching a peak point, the water level is reduced with the increase of the steam flow. This increase in the initial time of water level is called the phenomenon of "false water level". The peak and the lag time of the "false water level" depend on the scale of the steam flow disturbance and the gradient of pressure. When there is a negative step change in steam flow, the water level shows an opposite way of change.

Therefore, with considering the safety and economy, we require the steam generator water level control system can be maintained at the water level setting value in the steady-state operating condition. And in the load transient, the system can be adjusted according to the load changes and keep the parameter in the vicinity of the predetermined range. Steam generator water level control system, as its main task, is to maintain the water level setting values within a certain margin of error in various operating conditions; in the case of turbine downtime, the system can automatically or manually adjust the water flow to bring the water level back to the set value range.

Due to the "shrink up" effect of the saturated water inside the steam generator , the steam generator's "false water level" phenomenon is supposed to be very serious in the condition that water flow, especially the steam flow have a change of large-scale. Most of the emergency shutdown events due to steam generator are caused by "false water level". In the survey of the pressurized water reactor plants in-service, it shows that more than 30% of the accident shutdown was due to the main water supply system incidents. In French pressurized water reactor nuclear station, about 30% of emergency shutdown relates to the negative control of the steam generator water level in a low-load) operating condition. Unexpected emergency shutdown will reduce the availability of the power plants, leading to huge economic losses.

Traditional nuclear power stations adopt conventional cascade PI (D) control method which can meet the demand of the steam generator water level control in high load operating conditions, but there are still some disadvantages: the steam generator water level is a highly complex nonlinear parameter, time-varying non-minimum phase system, with a small stability margin. To obtain a good control effect in the case of variable operating conditions, it often requires changes in the parameters and gain of PI (D) controller. As the PI (D) controller's parameters of the analog control system are difficult to achieve setting online and the gain cannot be adjusted too high, otherwise it would cause excessive control system response which will influence the valve's life. In addition, with the rapid development of domestic nuclear power plants, new kinds of nuclear power plants are supposed to be not only with a grid base load, but also able to meet the requirement of peak regulation. Therefore, the conventional PI (D) control method is difficult to meet the full process control   requirements of modern nuclear power plants.

2   Models of SG Water Level Control System

Currently the most widely used steam generator water level control system is a three impulse PID control system composed of water flow, steam flow and steam generator water level [1].

(1) Affect on the water level with the main feed water flow

If the water temperature is lower than the boiling water temperature in the evaporator, it would lead to a reduction of water bubbles. In other words, when the water flow increases, there won't be an immediate increase in water level, which led to a reverse response of the system. This phenomenon is similar to the situation when the water flow rate decreases, SG water level has also given the reverse change. Practical experience and observation shows it exits a delay between changes in water flow rate and the actual water level. Considering the evaporator's volume and the dynamic hysteresis in dynamic response of water level and water flow integrative, the lower the temperature of water which goes into the evaporator the more evidently the control method is affected.

So, the transform function can be described as follows:

$$G_1(s) = \frac{H(s)}{Q_w(s)} = \frac{k_p e^{-\tau s}}{s(T_1 s + 1)(T_v s + 1)}$$

(1)

(2) Affect on the water level with the steam flow

The steam flow rate on the impact of SG water level with the water flow rate on the impact of SG water level is in a similar way but an opposite result. As the evaporator pressure changes with fluctuations in power, when a sudden change of power occurs, the evaporator pressure will also change. The contraction / expansion effect of water level have a greater impact on the water level than that of the changes of the water flow rate. Rapid contraction / expansion cause significant water level changes.

(3) SG Water level controller

The SG water level controller is a PID controller composed by a PI unit and a D (differential) unit. The transform function can be described as follows:

$$G(s) = K_{30}(1 + \frac{1}{\tau_{31} s})(\frac{1 + \tau_{36} s}{1 + 0.1\tau_{36} s})$$

(2)

(4) Feed water flow controller

The feed water flow controller is a PI controller. The transform function can be described as follows:

$$G(s) = K_{31}(1 + \frac{1}{\tau_{33} s})$$

(3)

Steam generator water level control system is consisting of the main channel, the bypass channel and the feed forward channel. The main channel water level setting unit generates a corresponding steam generator water level setting according to the secondary circuit load, and then the setting value of water level is compared with the measured value, resulting in a water level deviation. After the water level deviation amended by the variable gain cell, a calculation is conducted in the water level controller (PID) and resulting in a water flow signal. After superimposing the water flow rate signal and the feed-forward signal, the main water control valve opening signal is formed by the operation of the flow controller (PI) , then it produces an analog valve opening signal through the manual / automatic control. This signal can control the valve drive mechanism to regulate the valve opening which can change the water flow rat and finally the water level of the steam generator.

(5) Mathematical model of SG water level

In order to confirm the responses of the SG water level (H, %) about the feed water flow (Q_e, Kg/s) and steam flow (Q_v, Kg/s), we should build the mathematical model of SG water level. According to the SG water level characteristic above, the transform function can be described as follows:

$$G(s) = \frac{K_1}{s}(Q_e - Q_v) + \frac{K_2}{1+\tau_0 s}Q_v - K_3 \frac{1+\beta\tau s}{\tau^2 s^2 + 2\varepsilon\tau s + 1}Q_e \qquad (4)$$

This model is a muti-linearity model, which composed by several linearity models. Each model is built by a certain power operation point. For example, the low-load (0-20%FP) SG level characteristic is delegated by the linearity models of 10%FP and 20%FP, and the high-load (20-100%FP) is delegated by the linearity models of 50%FP and 100%FP. In fact, these power operation points are the typical operation conditions of nuclear power plants.

The parameters of formula (4) are as Table 1.

Table 1. The parameters for SG water level model

Power Value	K_1	K_2	K_3	τ_0	τ	ε	β
10% FP	0.0031	0.402	0.166	10	19.7	0.65	-0.08
20% FP	0.0035	0.339	0.207	10.4	12.5	1.6	0.44
50% FP	0.0035	0.188	0.055	6.4	13.3	0.62	0.20
100% FP	0.0035	0.131	0.028	4.7	6.6	1.68	0.20

The steam generator has a complex internal structure itself and there are quite a lot of parameters which affect the steam generator water level, so it is difficult to use a simple mathematical expression to describe the phenomenon of "false water level" caused by changes of water flow and steam flow. Moreover, with the operating conditions' differing, the parameters related to steam generator water level are also changing. The steam generator water level model has a strong property of time-varying, but in order to facilitate the study of the problem, a common-used simplified mathematical model of steam generator water level is adopted in this paper.

3   Simulation of Water Level Dynamic Characteristic

In order to verify the rationality of the models, the simulation of the water level is provided [2], [3]. The simulation experiment of the dynamic characteristics is based on MATLAB/Simulink as Fig.1.

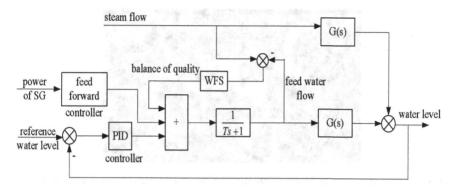

Fig. 1. The simulation model based on MATLAB/Simulink

The SG water level control system is applied on the nuclear power plant simulator in Wuhan University. The simulation model is established by the graphical modeling tools of RINSIM platform.

The nuclear power plant simulator can simulate some important equipments of PWR by using full simulation of 1:1 fidelity. The simulator is designed by three loops. The pressure differences between the steam pipes and feed water pipes are controlled by the speed of feed water pumps. So we can draw the change curves of each loop by the SIMCURV tool of the simulator in the conditions of different output loads [4].

When the output load is step decreased from 980MW to 750MW, the dynamic characteristics of SG water level are described as Fig.2 and the dynamic characteristics of feed water flow and steam flow are described as Fig.3.

In Fig.2, the SG water level control system can adapt the changes of load. The fluctuate of water level is small and the regulation time is short. Finally, the water level can stabilize on the reference water level (50%). In the condition of full-load, the experiments of step increasing and decreasing of feed water flow are described as Fig.4 and Fig.5 separately.

By the simulation experiments of the SG water level control system, the models can illuminate the changes of real SG water level in the conditions of different loads.

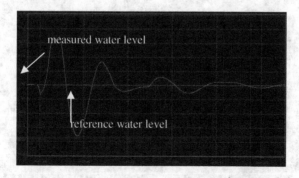

Fig. 2. The SG water level curves as step load reduction of turbine from 980MW to 750MW

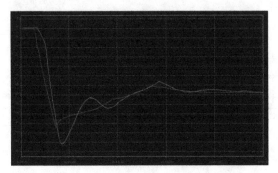

Fig. 3. The feed water flow and steam flow curves as step load reduction of turbine from 980MW to 750MW

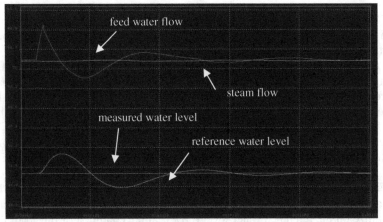

Fig. 4. The SG water level, feed water flow and steam flow curves as step increasing of feed water flow in full-**load** conditions

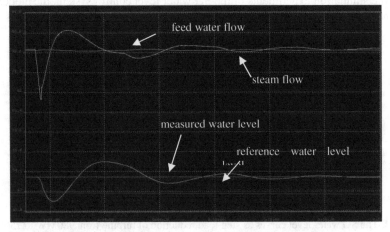

Fig. 5. The SG **water** level, feed water flow and steam flow curves

4 Conclusion

In this paper, the mathematical models of SG water level control system are built by MATLAB/Simulink, the simulation research based on the MATLAB/Simulink models is conducted, too. Then, based on RINSIM simulation platform of nuclear power plant simulator in Wuhan University, a simulation model on the SG level control system is established. By using the model, transient simulation experiments and researches with different conditions are conducted. Contrast with the MATLAB/Simulink simulation models, the good preciseness and identification performance of the RINSIM models are verified.

References

1. Irving, E., Miossec, C., Tassart, J.: Towards Efficient Full Automatic Operation of the PWR Steam Generator with Water Level Adaptive Control. In: Proceedings of Conference on Boiler Dynamics and Control in Nuclear Power Stations, pp. 309–329 (1998)
2. Habibiyan, H.: A Fuzzy-gain-scheduled Neural Controller for Nuclear Steam Generators. Annals of Nuclear Energy 31, 1765–1781 (2004)
3. Sudath, R., Munasinghe: Adaptive Neurofuzzy Controller to Regulate UTSG Water Level in Nuclear Power Plants. Nuclear Science 52, 421–429 (2005)
4. Qian, H., Li, C.: Full-load Water Level Digital Control System of Steam Generator. Journal of Shanghai University of Electric Power 25, 313–317 (2009)

Stabilization for Networked Control Systems with Packet Dropout Based on Average Dwell Time Method

Jinxia Xie[1], Yang Song[1,2,*], Xiaomin Tu[1], and Minrui Fei[1,2]

[1] School of Mechatronics Engineering and Automation, Shanghai University,
Shanghai, 200072
[2] Shanghai Key Laboratory of Power Station Automation Technology, Shanghai,
200072, P.R. China
Y_song@shu.edu.cn

Abstract. This paper proposes a new stabilization method for network control systems with stochastic packet dropout and network-induced delays. In terms of stochastic packet dropout, the NCS is modeled as Bernoulli process with two modes. By using average dwell time method, the sufficient conditions of NCS and the state feedback controller are derived.

Keywords: Network Control System, Bernoulli Process, Average Dwell Time, Exponential Stabilizable.

1 Introduction

With the rapid development and increasingly perfect of the computer networks technology, a new type distributed intelligent control system called-networked control system appears as the times require[4]. The primary advantages of NCS are source enjoyed together, ease of system diagnosis and maintenance. However, because of the existence of the network, the network-induced time delays that is often uncertain and time-varying[6][7], transmitted data error or data packet dropout makes the NCS is no longer a constant and certainty, which make the new control theories and methods are eagerly for investigation[5].

In the design procedure of the control system, different methods applied to ensure the stabilization of systems have very important sense[3]. Zhang and Yu using the a newly approach and the average dwell time approach investigated the exponential stabilization of the closed-loop NCS and gave the sufficient conditions with LMIs[1].

In this paper, by using average dwell time method, the modeling of NCSs with network-induced time delay and Bernoulli packet-dropout process will be established as switched systems, sufficient conditions and controller design to maintain the stabilization of the whole systems are derived. The acquired results have more universal significance.

2 Problem Formulation and Preliminaries

Assume the network between the plant and controller is illustrated in Figure.1, the time-driven sensor has the sampling period T, and the controller is event-driven.

---

* Corresponding Author.

K. Li et al. (Eds.): LSMS / ICSEE 2010, Part I, CCIS 97, pp. 8–13, 2010.

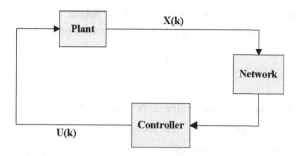

Fig. 1. Diagram of Networked Control Systems

Consider a linear system [2] [8]:

$$\begin{cases} \dot{x}(t) = Ax(t) + Bu(t) \\ \quad y(t) = Cx(t) \end{cases} \tag{1}$$

Where $x(t) \in R^n$ is the state, $u(t) \in R^n$ is control input, $y(t) \in R^n$ is the output, A, B, C are constant matrices with appropriate dimensions. Assume (A, B) and (C, A) are controllable and observable, respectively. Define network-induced time delay is $\tau_k < T$, can be described by $\tau_k = T/2 + \tau_k' < T$, Where $-T/2 < \tau_k' < T/2$.

Thus, above system can be described as follows:

$$x(k+1) = \Phi x(k) + \Gamma_0(\tau_k) u(k) + \Gamma_1(\tau_k) u(k-1) \tag{2}$$

Where $\Phi = e^{AT}$, denote $\tilde{F}(\tau_k') = \int_0^{\tau_k'} e^{As} ds, \Gamma_0 = \int_0^{T/2} e^{As} ds \cdot B, \Gamma_1 = \int_{T/2}^{T} e^{As} ds \cdot B,$

$$\delta = \left\| \int_0^{T/2} e^{As} ds \right\|_2, D = \delta e^{A(T/2)}, E_0 = B = -E_1, F(\tau_k') = \delta^{-1} \tilde{F}(\tau_k').$$

Therefore, the coefficient of system (2) can be get:

$$\Gamma_0(\tau_k) = \Gamma_0 + DF(\tau_k') E_0, \ \Gamma_1(\tau_k) = \Gamma_1 + DF(\tau_k') E_1.$$

Where $F(\tau_k')$ satisfies $F^T(\tau_k') F(\tau_k') \leq I$. The following omit τ_k' in $F(\tau_k')$ for simplicity. Denote augmented vector $Z(k) = \begin{bmatrix} x^T(k) & u^T(k-1) \end{bmatrix}^T$.

If the state feedback controller in networked control system is $u(k) = KZ(k)$, so the corresponding system described by:

$$Z(k+1) = \tilde{\Phi}_1 Z(k) = \left[\hat{\Phi}_1 + \hat{\Gamma}_1 K + \hat{D}F(\hat{E}_1 + E_0 K) \right] Z(k) \tag{3}$$

Where $\hat{\Phi}_1 = \begin{bmatrix} \Phi & \Gamma_1 \\ 0 & 0 \end{bmatrix}, \hat{\Gamma}_1 = \begin{bmatrix} \Gamma_0 \\ I \end{bmatrix}, \hat{D} = \begin{bmatrix} D \\ 0 \end{bmatrix}, \hat{E}_1 = \begin{bmatrix} 0 & E_1 \end{bmatrix}.$

When a packet dropout occurs, and suppose that the output of the controller is held at the previous value that $u(k) = u(k-1)$, thus the dynamics of the NCS described by:

$$Z(k+1) = \tilde{\Phi}_2 Z(k) = \begin{bmatrix} \Phi & \Gamma_1 + \Gamma_0 \\ 0 & I \end{bmatrix} Z(k) \tag{4}$$

Denote α_k is Bernoulli process represents packet dropout occurs in the network where $\alpha_k = 0$ implies that no packet dropout happens in the system, while $\alpha_k = 1$ implies packet dropout occurs. Assume $E(\alpha_k) = \alpha$:

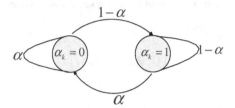

Fig. 2. Transition diagram of the NCS with packöet dropout

So the dynamics of the NCS can be described by the following two subsystems:

$$Z(k+1) = \tilde{\Phi}_{\sigma(k)} Z(k), \sigma(k) \in S = \{1,2\} \tag{5}$$

Definition 1. For any given initial conditions $(k_0, \phi) \in R^+ \times C^n$, (5) is globally exponentially stable if the solutions of (5) satisfy

$$\|x(k)\| \le c\lambda^{-(k-k_0)} \|x(k_0)\|, \forall k \ge k_0$$

Where $c > 0$ is a constant and $\lambda > 1$ is the decay rate.

3 Main Results

Theorem 1. For given positive scalars $\lambda_i, i \in \{1,2\}$ and $0 < \alpha < 1$ that $\lambda_1^\alpha \lambda_2^{(1-\alpha)} > \lambda > 1$, if exists symmetric positive-definite matrices X_1, X_2, matric Y ,positive scalars $\mu > 1, \xi > 0$ such that the following LMIs hold, moreover ,the gain matric of the state feedback controller is $Y = KX_1$,

$$\begin{bmatrix} -X_1 & * & * \\ \lambda_1 \hat{\Phi}_1 X_1 + \lambda_1 \hat{\Gamma}_1 Y & -X_1 + \lambda_1^2 \xi \hat{D} \hat{D}^T & * \\ \hat{E}_1 X_1 + E_0 Y & 0 & -\xi I \end{bmatrix} < 0 \tag{6}$$

$$\begin{bmatrix} -X_2 & \lambda_2 X_2 \tilde{\Phi}_2{}^T \\ \lambda_2 \tilde{\Phi}_2 X_2 & -X_2 \end{bmatrix} < 0 \tag{7}$$

$$X_1 \le \mu X_1 X_2^{-1} X_1, X_2 \le \mu X_2 X_1^{-1} X_2 \tag{8}$$

$$T_a > T_a' = \frac{\ln \mu}{2 \ln \lambda} \tag{9}$$

then (5) is globally exponentially stabilizable with a decay rate λ^ρ, $\rho = 1 + (\ln \mu / -2 T_a \ln \lambda)$, where T_a is the average dwell time.

Proof: choose the following piecewise Lyapunov function for system (5)[6]:

$$V_{\sigma(k)}(k) = Z^T(k) P_{\sigma(k)} Z(k)$$

Define $Z(k) = \lambda_i^{-(k-k_0)} \eta(k)$, we obtain

$$\eta(k+1) = \lambda_i \tilde{\Phi}_i \eta(k) \tag{10}$$

Choose the Lyapunov function $W_i(k) = \eta^T(k) P_i \eta(k)$ for subsystem (10), then the forward difference for $W_i(k)$ along any trajectory of subsystem (10) is given by

$$\Delta W_i(k) = \eta(k)^T \Omega_i \eta(k), \text{ where } \Omega_i = \lambda_i^2 \tilde{\Phi}_i{}^T P_i \tilde{\Phi}_i - P_i.$$

Combine (6) with $Y = KX_1$, denote $X_1^{-1} = P_1$, we can get:

$$\begin{bmatrix} -P_1^{-1} & * & * \\ \lambda_1 \hat{\Phi}_1 P_1^{-1} + \lambda_1 \hat{\Gamma}_1 K P_1^{-1} & -P_1^{-1} + \lambda_1^2 \xi \hat{D} \hat{D}^T & * \\ \hat{E}_1 P_1^{-1} + E_0 K P_1^{-1} & 0 & -\xi I \end{bmatrix} < 0$$

Above inequality multiplied by $diag\{P_1, P_1, I\}$ on both sides, and applying schur's complement, it follows that

$$\begin{bmatrix} -P_1 & * \\ \lambda_1 P_1 (\hat{\Phi}_1 + \hat{\Gamma}_1 K) & -P_1 \end{bmatrix} + \xi \begin{bmatrix} 0 \\ \lambda_1 P_1 \hat{D} \end{bmatrix} \begin{bmatrix} 0 \\ \lambda_1 P_1 \hat{D} \end{bmatrix}^T + \xi^{-1} \begin{bmatrix} (\hat{E}_1 + E_0 K)^T \\ 0 \end{bmatrix} \begin{bmatrix} (\hat{E}_1 + E_0 K) & 0 \end{bmatrix} < 0$$

Then, we can get:

$$\begin{bmatrix} -P_1 & * \\ \lambda_1 P_1 (\hat{\Phi}_1 + \hat{\Gamma}_1 K) & -P_1 \end{bmatrix} + \begin{bmatrix} 0 \\ \lambda_1 P_1 \hat{D} \end{bmatrix} F \begin{bmatrix} (\hat{E}_1 + E_0 K) & 0 \end{bmatrix} + \begin{bmatrix} (\hat{E}_1 + E_0 K)^T \\ 0 \end{bmatrix} F^T \begin{bmatrix} 0 \\ \lambda_1 P_1 \hat{D} \end{bmatrix}^T < 0$$

we can further obtain and by schur's complement: $\Omega_1 = \lambda_1^2 \tilde{\Phi}_1{}^T P_1 \tilde{\Phi}_1 - P_1 < 0$.

We also can get by (7) with $X_2^{-1} = P_2 : \Omega_2 = \lambda_2^2 \tilde{\Phi}_2^T P_2 \tilde{\Phi}_2 - P_2 < 0$.

For nonzero $\eta(k)$, $\Omega_i < 0$, $i \in \{1,2\}$, it is easy to obtain $W_i(k) < W_i(k_0)$, So that:

$$V_i(k) = \lambda_i^{-2(k-k_0)} \eta^T(k) P_i \eta(k) = \lambda_i^{-2(k-k_0)} W_i(k) < \lambda_i^{-2(k-k_0)} W_i(k_0) = \lambda_i^{-2(k-k_0)} V_i(k_0)$$

Therefore, by it follows that $\Omega_i < 0$, $i \in \{1,2\}$ is equivalent to (6)-(7).

Then, by (8) and expression of $V_{\sigma(k)}(k)$ such that

$$V_{\sigma(k)}(k) = Z^T(k) P_{\sigma(k)} Z(k) \leq Z^T(k) \mu P_{\sigma(k-1)} Z(k) = \mu V_{\sigma(k-1)}(k)$$

It is easy to see that

$$V_{\sigma(k)}(k) \leq \lambda_{\sigma(k_i)}^{-2(k-k_i)} V_{\sigma(k_i)}(k_i) \leq \lambda_{\sigma(k_i)}^{-2(k-k_i)} \mu V_{\sigma(k_{i-1})}(k_i) \leq \lambda_{\sigma(k_i)}^{-2(k-k_i)} \mu \lambda_{\sigma(k_{i-1})}^{-2(k_i-k_{i-1})} V_{\sigma(k_{i-1})}(k_{i-1}$$

...   (11)

$$\leq \mu^{N_\sigma[0,k)} \left(\lambda_1^\alpha \lambda_2^{(1-\alpha)} \right)^{-2(k-k_0)} V_{\sigma(0)}(k_0) < \mu^{N_\sigma[0,k)} \lambda^{-2(k-k_0)} V_{\sigma(0)}(k_0)$$

We can further obtain

$$V_{\sigma(k)}(k) < \mu^{N_\sigma[0,k)} \lambda^{-2(k-k_0)} V_{\sigma(0)}(k_0) = \lambda^{N_\sigma[0,k)(\ln \mu / \ln \lambda)} \lambda^{-2(k-k_0)} V_{\sigma(0)}(k_0)$$

$$= \lambda^{-2(k-k_0)(\ln \mu / -2T_a \ln \lambda)} \lambda^{-2(k-k_0)} V_{\sigma(0)}(k_0) = \lambda^{-2(k-k_0)(1+(\ln \mu / -2T_a \ln \lambda))} V_{\sigma(0)}(k_0) \quad (12)$$

$$= \left[\lambda^\rho \right]^{-2(k-k_0)} V_{\sigma(0)}(k_0)$$

Where $\rho = 1 + (\ln \mu / -2T_a \ln \lambda)$. We have:

$$a \|x(k)\|^2 \leq V_{\sigma(k)}(k) < \left[\lambda^\rho \right]^{-2(k-k_0)} V_{\sigma(0)}(k_0) \leq \left[\lambda^\rho \right]^{-2(k-k_0)} b \|x(k_0)\|^2$$

By simple computation, that

$$\|x(k)\| \leq \sqrt{b/a} \left[\lambda^\rho \right]^{-2(k-k_0)} \|x(k_0)\| \quad (13)$$

Where $a = \min_{i \in \{1,2\}} \lambda_{\min}(P_i)$, $b = \max_{i \in \{1,2\}} \lambda_{\max}(P_i)$, and $\lambda_{\min}(P_i)$ and $\lambda_{\max}(P_i)$ are the minimum and maximum eigenvalues of P_i respectively. For any switching signal with $T_a > T_a' = \dfrac{\ln \mu}{2 \ln \lambda}$, we have $\lambda > 1$ that guarantee that $\lambda^\rho > 1$. Hence, (5) with a decay rate λ^ρ is globally exponentially stabilizable by definition 1, this theorem holds.

4   Conclusions

In this paper, based on Bernoulli characteristics of the data-packets dropout, a new approach has been presented to study the stabilization of the NCS. Firstly, modeling the switched system with two subsystems for NCSs is established, and only considered the network in the NCSs between the plant and the state-feedback controller. Then under Lyapunov theory and average dwell time method, sufficient conditions and state feedback controller design for the stabilization of the NCS with a given decay degree via LMI formulations have been put forward.

Acknowledgment. This work is supported by National Natural Science Fund (60904016, 60774059, 60974097). Shanghai University Innovation Fund for Postgraduates(SHUCX102217). Excellent Youth Scholar of high education of Shanghai, and Shanghai University "11th Five-Year Plan" 211 Construction Project.

References

1. Zhang, W.A., Yu, L.: Output Feedback Stabilization of Networked Control Systems with Packet Dropouts. J. IEEE Trans. on Automatic Control 52(9), 1705–1710 (2007)
2. Ma, W.G., Cheng, S.: Stochastic Stability for Networked Control Systems. J. Acta Automatica Sinica 33(8), 878–882 (2007) (in Chinese)
3. Xie, D., Wang, Q., Wu, Y.: Average Dwell Time Approach To l_2 Gain Control Synthesis of Switched Linear Systems with Time Delay in Detection of Switching Signal. J. IET Control Theory and Applications 3(6), 763–771 (2009)
4. Zhang, W.A., Yu, L.: New Approach to Stabilization of Networked Control Systems With Time-Varying Delays. J. IET Control Theory and Applications 2(12), 1094–1104 (2008)
5. Song, Y., Fan, J., Fei, M.R., Yang, T.C.: Robust H $_\infty$ Control of Discrete Switched System with Time Delay. J. Applied Mathematics and Computation 205, 159–169 (2008)
6. Zhang, W.A., Yu, L.: A Robust Control Approach to Stabilization of Networked Control Systems with Time-Varying Delays. J. Automatica 45, 2440–2445 (2009)
7. Pan, S.G., Sun, J., Zhao, S.W.: Stabilization of Discrete-Time Markovian Jump Linear Systems via Time-delayed and Implulsive Controllers. J. Automatica 44, 2954–2958 (2008)
8. Ma, W.G., Cheng, S.: State Feedback Guaranteed Cost of Networked Systems. J. Information and Control 36(3), 340–351 (2007) (in Chinese)

Modeling of Real-Time Double Loops System in Predicting Sintering's BTP

Wushan Cheng

Shanghai University of Engineering Science, Shanghai 200065
cwushan@163.com

Abstract. In this paper, a double loops system, which based on the property of the large delay and time-varying of sintering process, is proposed to solve a challenging problem for building a system model of dynamic vary structure and vary weights from the given input and output data to predict the burning through point (*BTP*). A position track fuzzy controller is used to adjust the speed of sinter in outer loop, and an optimum Self-organizing Genetic Algorithms Neural Networks is also presented. The comparison of the actual process and the simulative process by OSGANN demonstrate that the performance and capability of the proposed system are superior.

Keywords: self-organized neural network, *BTP*, double-loop control, sinter production, integrate intelligent system.

1 Introduction

Sintering is the most widely used agglomeration process for iron ores and is a very important chain of iron making. In general, the process of sintering includes three major phases: first, it involves blending all the ores thoroughly according to certain proportions and adding water to the ore mix to produce particles, secondly, the actual sintering operation is initiated by the ignition hood, and finally, the finished sinter is broken up, cooled, and screened. During the three phases, the control of burn-through point goes through the whole sintering production, and that is one of the major means which affects the quality of sintering. Due to the complexity, time-varying and large delay for *BTP* control system, a great number of technologists and control experts devoted themselves in this research area in the recent 20 years, and the result of the research and development can be classified into the following three categories.

1.1 Mathematical Model

According to the chemical/physical characteristics for sintering, a model were formulated as a series of differential equation to describe the relation between the thick material, ignition temperature and the bellow's temperature at the tail of the machine[1]. For the time-varying and the random of the sintering process, many mechanisms have still not been understood. The dynamic model is tenable at a certain boundary condition. But it is difficult to cover the whole process by a certain mathematical model[2][3].

K. Li et al. (Eds.): LSMS / ICSEE 2010, Part I, CCIS 97, pp. 14–24, 2010.

1.2 Genetic Neural Networks

For the fast approaching of neural network, a model can be established rapidly from the given input and output data[4]. And it can also solve the problem of this long time delay system. Genetic algorithm is used to optimize the parameters of the network and improve the generalization of the system, but it hasn't still report to be used in real-time control as yet.

1.3 Rule-Based Model

The rule-base, acknowledge, data-base and inference mechanism can be constructed by the technologist and operation experts' experience[5], Rule base and inject machine are mainly used in estimating the process, analyzing cause and deciding guideline. Acknowledge includes operation data, fact, mathematical model and elicitation and unit knowledge, data-base stores real-time data from production and equipment. Unfortunately, most results of this model are still simulation results.

2 Integrate Intelligent System

According to above causes, the integrated intelligent control system for sintering process is led by combing the process control with technology reform in last 10 years. In this system, a double loops system is acted as predictor, the sintering status after 35 minutes at the end of the sintering machine can be predicted in the light of the operating data at the head of the machine, the predicting results will be returned directly to adjust the parameters for fuzzy controller's membership function in order to realize the mini-adjust for machine's velocity in advance. Considering the time-varying and disturbances of the sintering process, the outer loop of the system with negative feedback way is used to ensure the system robustness. The control frame is shown in figure 1.

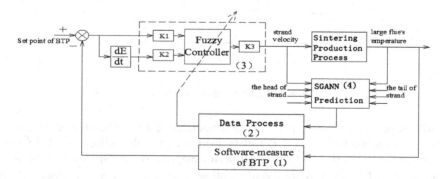

Fig. 1. Schematic diagram of the real-time double loops system

1. The software-measure method can be used to realize the real-time measurement of BTP.

2. The λ equals the average process value of 4 output parameters of SGANN with different weight, which is 18# bellow temperature,18# bellow pressure, the large flue press and the large flue pressure respectively.

3. The output of FC will be shifted with the membership functionμ,and the variable value of μis only coming from the λ.

4. Self-organizing Competitive Neural Networks with genetic algorithm can be trained as predictor, which training samples come from one third of sintering data of the latest 2 years, the remain data is selected as the operating parameter of SGAFC in real time.

2.1 A Soft-Measure Model of *BTP*

According to sintering theory, along with the sintering direction, the temperature of the bed of material becomes gradually higher, combustion zone is drift towards thicken, and the temperature of the bellows will not decrease until the whole process completed. So the position of BTP expression is the peak of the bellow's temperature. In the practical sintering process, the temperature of each bellow will form like curve as figure 2. There can be an approximately conic relationship between bellow's temperature near the end of the sintering and the position of the bellow.

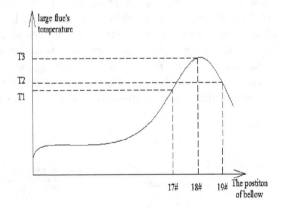

Fig. 2. The temperature of bollows

BTP is just the inflecting point of the temperature curve. When the practice distance between two bellows is 3.95 meters, the position of the bellow $17^{#} 18^{#} 19^{#}$ can be denoted as *-3.95, 0, 3.95* by the coordinate, temperature corresponding to the bellows are T_1、T_2、T_3 respectively. It's nonlinear regression can be as follows:

$$T_k = AX_k^2 + BX_k + C \qquad (1)$$

Where X_k represents the position of the bellow, T_k is the temperature of the bellow, X_k A, B and C are undetermined coefficient respectively. The fix relation can be obtained from this equation.

$$X_{max} = -\frac{B}{2A} \qquad (2)$$

At the max temperature point T_{max} of the bellow is in correspondence with the distance of deviation between the two bellows, which predicates the final position is shown as follows.

$$L_{BTP0} = L_{18} + X_{max} \qquad (3)$$

Where L_{BTP0} is the normal position of the burning through point, L_{18} is the position of the bellow *18*, X_{max} is the distance of deviation between the two bellows.

As the exhaust gas temperature of large flue is a monotonous increase progressively, whose change tendency will affect the whole sintering process. The temperature compensation will be corrected the curve through temperature of the large flue. In this paper, the liner regression relation between the last two year's temperature of large flue and blower number is used to simulate the liner relationship:

$$y = -1.772x + 147.2 \qquad (4)$$

Where y is temperature of the large flues, x is the position of the bellow temperature.

According to operating experience, when the temperature error ΔT_f between the real large flue's temperature and the normal temperature can be linear drafts, the revising formula is:

$$L_{BTP} = L_{BTP0} \qquad when\ 143°C < T_f < 147° \qquad (5)$$

$$L_{BTP} = L_{BTP0} + \alpha \Delta T_f \qquad otherwise$$

Among them, α is close through weight, about equals -0.56.

2.2 Fuzzy Controller

The fuzzy controller is the main controller of the system, which contains two parts, the first part contains the position error $e(k)$ and the position change rate $e_c(k)$, the second part is the strand velocity, which is closest equivalent to the gauss function of the input.

$$\mu = e^{-\frac{(x-m)^2}{\lambda^2}} \qquad (6)$$

Where μ is the membership function of the system, m is the average value of the function, and λ is the mean square error which coming from the data process unit.

According to the theory of fuzzy control, if X_i is input, the demand control output will be $U_i (i = 1,2,...n)$, but the controlled result actually will be changed instead of be exact U_i , by using R to represent its fuzzy relation, all the rules of the relation have intersections, the following equation is existence.

$$\mathop{\exists}_{1\leq i\leq n}\mathop{\exists}_{u\in U}(X_i\circ R)(u)\neq U_i \tag{7}$$

Where "∘" is defined as compact support, that is a compound relation. For the intersecting membership function, the following equation can be obtained.

$$\mathop{\exists}_{\substack{1\leq i\leq n\\ j\neq i}} X_i\cap X_j \neq \Phi \tag{8}$$

$X_j\ (k\neq i, k\in\{1,2,\cdots,n\})$ is exist and the rules have interaction.

Proving as follow.

$$X_j\cap X_i \neq \Phi, i, j = 1,2,3,...,n \tag{9}$$

and $i\neq j$ except $j = k$, the X_k is interest with

$$\mathop{\vee}_{x\in X}[X_k\wedge X_i(x)]=\alpha, 0\leq \alpha \leq 1 \tag{10}$$

Therefore

$$(X_i\circ R)(u)= \mathop{\vee}_{x\in X} X_i(x)\wedge R(x,u)= \mathop{\vee}_{x\in X}\left\{X_i(x)\wedge\left[\mathop{\vee}_{j=1}^{n}(X_j(x)\wedge U_j(u))\right]\right\}=$$

$$\mathop{\vee}_{x\in X}\left\{\mathop{\vee}_{j=1}^{n}[X_i(x)\wedge X_j(x)\wedge U_j(u)]\right\}=$$

$$\mathop{\vee}_{x\in X}[X_i(x)\wedge X_i(x)\wedge U_i(u)]\mathop{\vee}_{x\in X}[X_i(x)\wedge X_k(x)\wedge U_k(u)]=$$

$$U_i(u)\vee[\alpha\wedge U_k(u)]\geq U_i(u) \tag{11}$$

Also means

$$X_i\circ R\supseteq U_i,\ X_i\circ R\neq U_i, \tag{12}$$

When the membership function of FC change, the outputs of FC will be shifted with the input very accordingly, though the input of FC don't very. Figure3 gives the relationship of the output of FC and the output λ of the data process.

In both the predicting of the NN, and the level and smooth processing of data process, whose output λ is a vary value dynamically. When the λ is very small, the system has a big overshooting and a fair-sized steady-state error. When the λ is very big, system haven't a overshooting, but have a response slowly. Only when λ is suitable value ($\lambda=1$), does the system have quick response speed, overshooting free and no steady-state error. so the rule has interaction

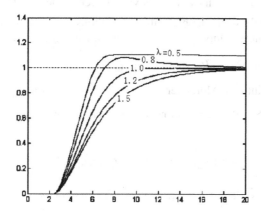

Fig. 3. the relationship of the output of FC and the λ

3 Optimum Self-organized Genetic Algorithms

Standard genetic algorithm uses a fixed selection rate, crossover rate and mutation rate, all these are prone to cause premature convergence. According to the preference of Sinter's complex and dynamical process, standard genetic algorithm is difficult enough to optimize the rules and membership function of Fuzzy Controller. In the early stage of evolution, the difference of the individual in population is large, so the selection and crossover play a good effect of evolution. In the later stage, population lost the variety of individual for the invariable probability of crossover and mutation, and cause premature convergence of the GA to a local optimum, the children are reproduced from the similar chromosomes. So, adaptive probabilities of crossover and mutation must be applied in the genetic algorithms for optimize the rules and membership function of fuzzy controller. Based on this aspect, Self-Organized Genetic Algorithms is used to solve the problem.

First, the crossover rate and mutation rate can be observed by the following form:

$$P_c = k_1 / (\bar{f} - f_{\min})$$
$$P_m = k_2 / (\bar{f} - f_{\min})$$

(13)

Where P_c and P_m do not depend on the fitness value of any particular solution. When a population converges to a global optimal solution, P_c and P_m will increase and may cause the disruption of the near-optimal solutions. The population may converge to the global optimum. Though the local optimum may be prevented, the performance of the GA will certainly deteriorate.

To overcome this problem, a 'good' solution of the population must be prevented. The normal way can be achieved by having lower values of P_c and P_m for low fitness solutions. Only while the high fitness solutions are used in the convergence of the GA, the low fitness solutions can be prevented from getting stuck at a local optimum. The value of P_c or P_m should depend not only on $\bar{f} - f_{min}$, but also on f' or f of the solutions. Moreover, the closer f' or f is to f_{min}, the smaller P_c or P_m should be. So, the expressions for P_c and P_m now take as the following forms.

$$P_c = k_1 (f_{min} - f') / (\bar{f} - f_{min}), k_1 \leq 1.0$$
$$P_m = k_2 (f - f_{min}) / (\bar{f} - f_{min}), k_2 \leq 1.0$$
(14)

The algorithm are constrain P_c and P_m to the range from 0.0 to 0.1

To prevent the overshooting of P_c and P_m beyond 1.0, the constraints are presented as follow.

$$P_c = k_3, f_{min} < f'$$
$$P_m = k_4, f < f_{min}$$
(15)

Where $k_3, k_4 \leq 1.0$. We saw that for a solution with the minimum fitness value, P_c and P_m are both zero. The best solution in a population is transferred undisrupted into the next generation. Together with the selection mechanism, this may lead to an exponential growth of the solution in the population and may premature convergence.

Usually, the choice of P_c and P_m are given as following for convenience:

$$P_c = \begin{cases} k_1 (f_{min} - f') / (\bar{f} - f_{min}), & f' \leq f_{min} \\ k_3, & f' > f_{min} \end{cases}$$
$$P_m = \begin{cases} k_2 (f_{min} - f') / (\bar{f} - f_{min}), & f' \leq f_{min} \\ k_4, & f' > f_{min} \end{cases}$$
(16)

Where k_1, k_2, k_3 *and* k_4 are less than 1, k_3 *and* k_4 are bigger.

Then, we consider it as a bad chromosome when the fitness is lower than the average fitness, or a good chromosome when the fitness is higher than the average fitness. And having lower values of P_c and P_m for high fitness solutions. But the rate of crossover and mutation are approach to zero when the fitness value of the individual is equal to the maximum fitness value of current generation. This approach adapt to the later stage of the evolution but not early stage. Because each individual has a good capability during the later stage and it cannot be mutated. And the good solution of the problem in early stage may not is the best in global area, so we shall solve the problem of slow convergent speed by a modified genetic algorithm.

$$
P_c = \begin{cases} P_{c1} - \dfrac{P_{c1} - P_{c2}}{f_{\max} - f_{avg}}(f' - f_{avg}), & f' > f_{avg} \\ P_{c1}, & f' < f_{avg} \end{cases}
$$

$$
P_m = \begin{cases} P_{m1} - \dfrac{P_{m1} - P_{m2}}{f_{\max} - f_{avg}}(f' - f_{avg}), & f' > f_{avg} \\ P_{m1}, & f' < f_{avg} \end{cases}
$$

(17)

The equation put some value of P_c and P_m, when the fitness value of the individual is equal to the maximum fitness value of current generation. But the f_{avg} of the population is not convergence to the best with the increasing of generation, and it may cause disruption.

At last, the optimum genetic algorithm is applied to a traditional Multilayer feed forward neural network to be simulated as the actual sintering process.

4 Simulation

In this section, three types of the simulation networks, which are BP Neural Network, Basic Linear Matrix Neural Network, Optimum Self-organizing Neural Networks respectively, are compared to illustrate the performance of the SGANN predictor. The program is coded and simulated by Matlab language. Results for real-time comparisons of the simple BPNN, LMNN and optimum OSGANN are given in Fig 4, 5 and 6 respectively.

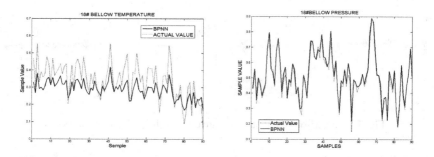

Fig. 4. Comparison of real-time performance of BPNN and actual data value

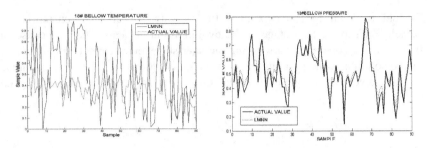

Fig. 5. Comparison of real-time performance of LMNN and actual data value

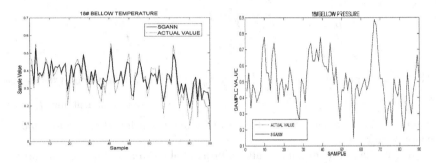

Fig. 6. Comparison of real-time performance of OSGANN and actual data value

From the Fig.4 to Fig. 6, the calculating value by OSGANN is most close to the actual value in the three types of the simulation networks. According to the above figures, the value of the pressure is better than the temperature which main because the reliability of the pressure sample is higher than the temperature, no matter the network architectures and weights.

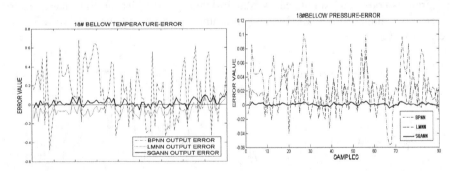

Fig. 7. Comparison of the mean square error of three kind of networks

In Fig.7, the test samples against real-time predicted value error for sintering predicting system can be reached by plotting LMNN, BPNN and SGANN model respectively, when the mean square error is given 0.001. According to the figure, only OSGANN can meet the requirement because the MSE is requested as 0.0001.

Table 1. Comparison of the MSE of three simulation networks

Sample	Sample#18	Sample#18	Sample#18
Network structure	BPNN	LMNN	SGANN
Network Parameters	4-15-4	4-15-4	4-15-4
Training time	3299 epochs	4163 epochs	5762 epochs
Mean square time(MSE)	0.000993389	0.000274878	0.000069812

As shown in Table 1, either LMNN or BPNN do apply a gradient descent algorithm in which the mean squared error between the network's output and the desired output is minimized. This creates a global cost function that is minimized iteratively by backpropagating the error from the output nodes to the input nodes. Once the network's error has decreased to less than or equal to the specified threshold, the network has converged and is considered to be trained.

From the above comparison, we find that not only the calculating value by OS-GANN is most close to the actual value but also the performance of mean square and real-time predicted error are improved by OSGANN. So OSGANN is very advantageous, credible and reliable.

5   Conclusion

It is very difficult to construct a fuzzy model from sintering production to deal with the large time delay and time-varying system. And it makes the system fall into confusion. System must have great capabilities in prediction in order to predict the change of input states. So a double-loop system should be used to resist the disturbance from the sintering production. The software-measure method can be also used to realize the real-time measurement of *BTP*. Meanwhile, the self-organizing Neural Networks with optimum genetic algorithm can be trained as predictor. The load quantity, air penetrability, strands velocity and ignition temperature curves can be made as predictive parameters to 4 inputs of 18# bellow temperature, 18# bellow pressure, 18# large flue pressure. And the simulation results show that the optimum SGANN has better performance than Basic LMNN and BP.

References

1. Dawson, P.R.: Recent development in iron ore sintering. Ironmaking Steelmaking 20(2), 135–159 (1993)
2. Fan, X.H.: Adance prediction models of sinter chemical composition. Sintering Pelletizing 18(4), 1–5 (1993)
3. Kim, Y.H., Kwon, W.H.: An application of min-max generalized predictive control to sintering processes. Control Engineering Practics 6, 999–1007 (1998)

4. Azeem, M.F., Hanmandlu, M.: Structure Identification of Generalized Adaptive Neuro-Fuzzy Inference Systems. IEEE Transactions on Fuzzy Systems 11(5) (2003)
5. Hai, X., Yong, M., Shi, L., Xiang, L.H.: A Method to Simplify Fuzzy System Based of the Overlap of Fuzzy Sets. Journal of Fudan University (Natural Science) 39(2) (2000)
6. Ming, F.Q., Tao, L., Hui, F.X., Tao, J.: Adaptive Prediction System of Sintering Through Point Based on Self-organize Artificial Neural Network 10(6) (2000)
7. Fortemps, P., Roubens, M.: Ranking and defuzzification methods based on area compensation. Fuzzy Sets and Systems 82, 319–330 (1996)
8. Cheng, W.H.: Intelligent Control of the Complex Technology Process Based on Adaptive Pattern Clustering and Feature Map. Journal of Mathematical Problems in Engineering, 1024–123X

An Orthogonal Curvature Fiber Bragg Grating Sensor Array for Shape Reconstruction[*]

Jincong Yi[1,2], Xiaojin Zhu[1,**], Linyong Shen[1], Bing Sun[1], and Lina Jiang[1]

[1] School of Mechatronics Engineering and Automation, Shanghai University, Shanghai, 200072, P.R.China
[2] College of Computer and Information Science, Fujian Agriculture and Forestry University, Fuzhou, 350002, P.R. China
mgzhuxj@shu.edu.cn

Abstract. An orthogonal curvature fiber Bragg grating (FBG) sensor array is introduced, and it can detect the deformation and vibration of flexible structures such as rod, keel, etc. The sensor array composed of 20 sensors which were averagely distributed on four optical fibers was mounted on the body of cylindrical shape memory alloy(SMA) substrate with staggered orthogonal arrangement. By the method of calibration, the relation coefficient between curvature and wavelength shift was obtained and the curvature of sensor was calculated accordingly, then, the space shape was reconstructed with the help of the space curve fitting method based on curvature information of discrete points. In this paper, the operation principle, the design, packaging, calibration of FBG sensor array and the method of experiment were expounded in detail. The experiment result shows that the reconstructed spatial shape is lively, thus indicates that the relevant method and technology are feasible and practicable.

Keywords: Fiber Bragg grating sensor array, orthogonal curvature, sensor package, shape reconstruction.

1 Introduction

With the development of space technology, large flexible space structures such as truss, keel-like structure and antenna structure and so on were widely applied in spacecraft. Because these large flexible space structures usually possess such features as low stiffness, low damping, low natural frequencies, they are apt to be deformed by the effect of external disturbances, causing the performance reduced even be damaged[1].There is strict request to monitor the deformation, internal stress and damages and to realize in-orbit autonomous real-time monitor and real-time active vibration control of the structure[2,3], thus help guarantee the spacecraft to run in a stable and healthy state.

In recent years, FBG sensor array as an advanced detection technology due to the well-known advantages such as electrically passive operation, immunity to EMI, high

---

[*] This research is supported by program of National Nature Science Foundation of China (No.90716027), and key program of Shanghai Municipal Education Commission (No.09ZZ88).
[**] Corresponding author.

K. Li et al. (Eds.): LSMS / ICSEE 2010, Part I, CCIS 97, pp. 25–31, 2010.
© Springer-Verlag Berlin Heidelberg 2010

sensitivity, compact size and easily multiplexed in a serial fashion along a single fiber[4] has attracted considerable interest to apply the orthogonal curvature-based FBG sensor array to real-time strain measurement during the spacecraft operation.

With the research thinking considered, the real-time active monitoring of the deformation and vibration of the structure can be achieved by using orthogonal curvature-based fiber Bragg grating (FBG) sensor array.

The basic idea presented in this paper is described as follows: first, implant the FBG sensor array with a orthogonal distributed method, then the wavelength is measured by the demodulation, thus the curvature obtained according to the relation between wavelength and curvature for a pure bending rod, finally, reconstruction of the bending shape of the flexible rod by means of reconstruction algorithm and visualization of the shape can be achieved. The paper is featured with the operation principle, design, packaging process. Meanwhile, calibration method and experiment are also discussed, and the experiment results show that the curvature detection method and the technology of visualizing monitor of shape deformation using orthogonal curvature-based FBG sensor array is practicable, and thus provides an effective method and technical support for the further study and exploration in the spacecraft application.

2 Principle of Obtaining Curvature

For our experiment, it is required to reconstruct the shape of pure bending flexible rod in real time based on curvature. Obviously, how to obtain the curvature data is the key to reconstruct shape. But the experiment data is the wavelength detected by a FBG sensor array and it must be converted to the curvature. According to the document[5], when a mechanical axial strain ε is applied to the grating and the temperature is stable, the shift in the reflected Bragg wavelength λ_B can be given by Eq.1:

$$\Delta \lambda_B = \lambda_B \cdot (1 - P_e) \cdot \varepsilon \tag{1}$$

Where P_e is the effective photo-elastic coefficient of the fiber. Generally, a silica core fiber has a photo-elastic constant of 0.22.

When a FBG sensor was stuck on the top of a bent beam as shown in Fig.1[6], the bend strain ε along the axial direction at any position on the surface of beam can be given by Eq.2 according to the pure bend theory.

$$\varepsilon = \frac{(\rho \pm y)d\theta - \rho d\theta}{\rho d\theta} = \pm \frac{y}{\rho} = \pm yk \tag{2}$$

Where k is the curvature of the bent beam, ρ is the curvature radius of the bended beam, $d\theta$ is the angle of the arc, and y is the distance from the neutral surface O_1O_2 of the beam to the sensing position, then Eq.3 can be deduced from Eq.1 and Eq.2.

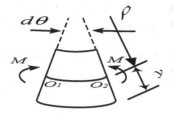

Fig. 1. The relation between curvature and bend strain

$$k = \frac{\Delta \lambda_B}{\lambda_B \cdot (1 - P_e) \cdot y} \qquad (3)$$

To each given coating condition, λ_B, P_e and y are constants, so k is linear to $\Delta \lambda_B$. Thus using Eq.3, it is easy to calculate the curvature k from the measured wavelength shift $\Delta \lambda_B$.

3 FBG Sensor Array Design and Packaging

With the development of the technology of wavelength-division-multiplexing (WDM) and spatial-division-multiplexing (SDM), it is feasible to design a sensor array. To design a FBG sensor array, some factors such as sensor number, interval, grating location, measurement range of instrument, packing and cost etc. must be taken into consideration. Because every device have more than one channel, it is easy to utilize each of the channels of demodulator to construct a SDM array. For WDM, it is required to determine the location and number of fiber Bragg grating in each channel[7]. In our experiment, four channels of the eight available monitoring channels of the demodulator were connected to four fiber-optic cables, which were arranged uniformly with 5 separate FBG sensors respectively. Four optical fibers were stuck evenly on the shape memory alloy (SMA) substrate in an orthogonal around way illustrated in Fig.2. In the figure, the internal circular cross section signifies the memory alloy substrate such as wire, rode, etc, the four solid small circles A, B and A', B' represent four optical fiber shown in Fig. 2(a), and the arrangement of the 5 fiber grating along the axial direction is shown as Fig. 2(b). A, B and A', B' are in pairs, and the crossed staggered arrangement can help increase the numbers of fiber grating within a limit space, thus improve the precision of the reconstruction of shape.

Once the sensor array was designed, the package must be followed. In order to obtain the linear relationship between the wavelength shift and the strain of the optical fiber, the shape memory alloy (SMA), possessed unique feature with being able to recover its original shape after deformed was used as a package substrate to fabricate

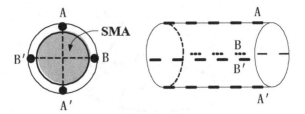

Fig. 2. The installation diagram of sensors measuring curvatures

the FBG sensor array[7,8]. In addition, to improve the tensile properties of the fiber grating sensors, shape memory alloy wire was pre-stressed too, and the pre-stress load device was introduced as shown in Fig.3. The packaged FBG sensor array is composed of four orthogonal optical fibers. The number of the grating sensor points is five in each fiber. Each of the five fiber grating is distributed in a uniform way. The diameter of each single fiber is 0.25mm and the length of the array is 2000mm.

Fig. 3. Pre-load device

4   Spatial Curve Reconstruction Algorithm

For a pure bending flexible rod, the bending shape of its central axis can approximately be regarded as a spatial curve, thus, the reconstruction of bending shape refers to determine the coordinate of a changing spatial curve. So the moving coordinate system concept is introduced. Assumed that the *1*st point O_1 of the curve as the origin of the coordinates system, the fixed coordinate system $\sigma^{(0)} = [O_1; x, y, z]$ and the moving coordinate system $\sigma^{(1)} = [O_1; a_1, b_1, c_1]$ are established as shown in Fig.4, where c_1, a_1 and b_1 axis are the tangential direction and the two orthogonal curvature directions of O_1 respectively, k_1 is the compound direction of the two orthogonal curvature directions. And the plane confirmed by k_1 and c_1 axis is just the osculating plane π_1 of micro segment O_1O_2 [9]. Similarly, other point's moving coordinate systems can be structured correspondingly. Thus, the micro arc segment O_1O_2 can be regarded as a micro curve on the osculating plane π_1 when the point O_2 is approaching to the point O_1 and the arc segment is extremely small without considering the twisted state. By this way, we can locate the point O_2 on the osculating plane π_1 according to the arc length ds and curvature value k_1. Ulteriorly, the coordinate value of point O_2 in the fixed coordinate system $\sigma^{(0)}$ is obtained. In this way, the space curve shape can be reconstructed by means of connecting every different radius micro arc segments which are on the different osculating plane.

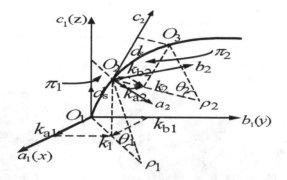

Fig. 4. Schematic diagram of the coordinates system

5 Calibration and Experiment

Generally, the relation between the wavelength shift of FBG sensor and the variation of detected physical parameter is not linear correspondence. So it is necessary to calibrate each FBG sensor to determine the exact relationship between the curvature and the wavelength shift of FBG sensor. For the orthogonal FBG sensor array, the calibration of each sensor must be carried out on the two perpendicular orthogonal planes as shown in Fig.5, and the relationship between the curvature and the wavelength shift of the *1*st sensor of the *1*st optical fiber is as shown in Fig. 6. There are five FBG sensors distributed on one optical fiber and four optical fiber 20 FBG sensors in total. In Fig. 6, the result shows that the relation between the curvature and the wavelength shift is good linear and verifies the effectiveness of the array package. By the same way, other FBG sensors can be calibrated and the corresponding curve can also be obtained respectively to be used for the shape reconstruction.

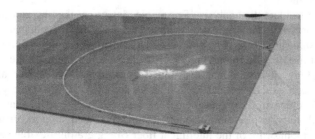

Fig. 5. FBG sensor array calibration with standard arc calibration board

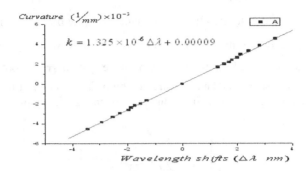

Fig. 6. Relation between curvature and wavelength shift of 1st optical fiber

Fig. 7. Schematic diagram of reconstruction of space shape

We reconstructed the shape according to the curvature with the help of the algorithm of spatial curve reconstruction mentioned above. The reconstructed shape were showed in Fig. 7(a), (b). It can be seen that the display on the screen of bending shapes of the rod are consistent with the actual bending shape and have real-time display properties, thus to demonstrate that the bending shape of a flexible rod based on FBG sensor array can be reconstructed lively.

6 Conclusion

This paper presented a countermeasure of developing a FBG sensor array under the background of active monitoring technology oriented to structural vibration of flexible aerospace vehicle.

The article considers the analysis of principle, design, packaging and calibration of the FBG sensor array and the experiments. The analysis of technical method and experiment result show that it is practicable to perceive the three-dimension space morphology of the flexible rod and reconstruct the bending shape on the screen based on the technology of orthogonal distributed embedded FBG sensor array. Using the method, it can exactly reflect the shape variation of the structure with a living display on the screen in the experiment. The research on the FBG sensor array can provide a good study, exploration idea and technical method support for the real-time perception and monitoring of the intelligent fiber Bragg grating flexible structure with a non-vision method.

Acknowledgment

This paper is sponsored by program of National Nature Science Foundation of China (90716027), key program of Shanghai Municipal Education Commission (09ZZ88), Mechatronics Engineering Innovation Group project from Shanghai Education Commission, Shanghai University "11th Five-Year Plan" 211 Construction Project and Shanghai Key Laboratory of Power Station Automation Technology.

References

1. Rapp, S., Kang, L., Han, J., Mueller, U., Baier, H.: Displacement Field Estimation for a Two-Dimensional Structure Using Fiber Bragg Grating Sensors. Smart Materials and Structures 18, 025006, 12 (2009)
2. Xiaojin, Z., Meiyu, L., Hao, C., Xiaoyu, Z.: 3D Reconstruction Method of Space Manipulator Vibration Shape Based on Structural Curvatures. In: 8th International Conference on Electronic Measurement and Instruments, pp. 2772–2775. IEEE Press, Xian (2007)
3. Kim, K., Kwon, W., Yoon, J., Lee, J., Hwang, Y.: Determination of Local Thrust Coefficients of an Aerospace Component With the Fibre Bragg Grating Array. Proceedings of the Institution of Mechanical Engineers, Part G: Journal of Aerospace Engineering 3, 331–338 (2010)
4. Tiwari, U., Mishra, V., Jain, S., Kesavan, K., Ravisankar, K., Singh, N., Poddar, G., Kapur, P.: Health Monitoring of Steel and Concrete Structures Using Fibre Bragg Grating Sensors. Current Science 11, 1539–1543 (2009)
5. Hill, K., Meltz, G.: Fiber Bragg Grating Technology Fundamentals and Overview. Lightwave Technology 8, 1263–1276 (1997)
6. Hongchao, F., Jinwu, Q., Yanan, Z., Linyong, S.: Vibration Intelligent Test of Large-scale Flexible Surface. Chinese Journal of Mechanical Engineering 7, 202–208 (2008)
7. Dulieu-Barton, J.M., Quinn, S., Ye, C.C.: Optical Fibre Bragg Grating Strain Sensors: Modern Stress and Strain Analysis. Eureka Magazine, 1–2 (2009)
8. Zhang, X., Li, X.: Design, Fabrication and Characterization of Optical Microring Sensors on Metal Substrates. Micromechanics and Microengineering 18, 015025–31 (2008)
9. Buqing, S.: Differential Geometry. Higher Education Press, Beijing (1979)

Implementation of the PCB Pattern Matching System to Detect Defects

Cheol-Hong Moon, Hyun-Chul Jang, and Jin-Kook Jun*

Gwangju University, *OKins Electronics Co., Korea
chmoon@gwangju.ac.kr, psycotist@nate.com, jkjun@okinc.co.kr

Abstract. FPGA-based PCB Pattern Matching System, which supports a Camera Link (Medium), was used to detect PCB defect patterns. For the automation of the vision inspection of the PCB production process, the system was optimized by implementing the vision library in IP, which is used to produce high speed processing FPGA-based systems and to detect defect patterns. The implemented IPs comprised of Pattern Matching IP, VGA Control IP, Memory Control IP and Single Clock Processing MAD Pattern Matching IP. Xilinx was used to process the image transmitted in high speed from Digital Camera, Vertex-4 type FPGA chip. It allowed the processing of 2352(H) * 1728(V) *8Bit image data transmitted from the camera without the need for a separate Frame Grabber Board[5] in the FPGA. In addition, it could check the image data on a PC. For pattern matching, it abstracted a 480*480 area out of the image, transmitted the image to each IP and displayed the Pattern Matching output result on a TFT-LCD.

Keywords: PCB, Pattern Matching, MAD, Single Clock Processing MAD Pattern Matching IP.

1 Introduction

Currently, the machine vision industry is focused on the development of ultra small digital devices with more functions to meet the needs of consumers. Therefore, technology to enhance the accumulation of PCB is needed. The spatial limitation had been resolved to some extent with the commercialization of m unit PCB. However, there are still many problems, such as high defect rate in the manufacturing process and in its structure [1][2]. X-rays have been used to detect the defects of PCB but the industry introduced Pattern Matching method using Machine Vision to make the detection more efficient and faster [7]. The Pattern Matching method is the most widely used image processing method in the machine vision industry. It became feasible with the rapid development of computer technology and the lower price. However, the calculation operations cannot be handled on a PC in real time if the image size is too large in the algorithm using Mean Absolute Difference (MAD), which counteracts the Normalized Gray-scale Correlation (NGC) [3][4]. In vision processing using Camera Link, an expensive Frame Grabber Board is needed to process the image data

K. Li et al. (Eds.): LSMS / ICSEE 2010, Part I, CCIS 97, pp. 32–38, 2009.

on PC, which is transmitted from a Digital Camera at high speed [5]. This article proposes a SoC type FPGA-based system. It supports the Medium of the Camera Link and allows real time pattern matching using the image transmitted from the Camera. Because it utilizes FPGA, the system can be compact and each IP designed by Logic can be reused in all other FPGA systems. The Logic IPs implemented in this article are Pattern Matching IPs using the MAD algorithm, VGA, and Memory Control IP.

2   MAD Pattern Matching

There are many different methods to find a similar image pattern using the given templates when an image is entered. Among them, the matching method, MAD subtracts the brightness value of the pixels corresponding to the overlaps between the template image and target image, and then adds the difference values. Equation 1 shows the MAD algorithm, which calculates the pixel difference between Template T and the overlapping part from the entered image I [3].

$$MAD = \frac{1}{MIN} \sum_{i=0}^{A-M} \sum_{j=0}^{B-N} \mid I(x_i)(y_j) - T(x_i)(y_j) \mid \tag{1}$$

To determine the absolute difference in the corresponding pixels, the entered image block moves the repeating area of (A - M)*(B - N) . For instance, if the actual template size is 64*64 and the target image is 480*480, the operation times will be (480-64)*(480-64), which is 173,000 times with the application of MAD. Moreover, in each inspection, the subtraction will be performed (64*64) times. The time complexity according to the calculations will increase if the size of the template and entered image becomes larger. Therefore, this article proposes a new method to reduce the subtraction operation times by processing 64 bytes in Multi as well as the operation time remarkably by processing MAD in parallel. Figure 1 is the block diagram of the MAD algorithm processing. It reads data from the target image at the size of the template and performs a MAD conformity test beginning with pixel 1 and repeating it at a line size (64pixel).

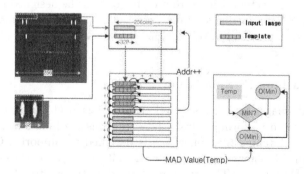

Fig. 1. MAD algorithm Block Diagram

3   PCB Pattern Matching IP System Design

When the comparison between the 64*64 size image of the template and 64*64 size image of the target image is completed, the Logic saves the final MAD Value. Simultaneously, it increases the address of the target image for the conformity test with new pixels and repeats the above process. When pattern matching of 480*480 area from the target image is completed, it retrieves the corresponding grid using the smallest Register value. Figure 2 shows the Pattern Matching System IP Configuration in this article. 4Tap image data transmitted from a digital camera is entered from Cam-Link-Top block in FPGA. It reconfigures it to the data format of Camera Link and begins transmission. It transmits the data to IP-Controller-Top. In VGA Control IP, it abstracts the 480*480 size area only from the image data and transmits them to a 7inch TFT-LCD.

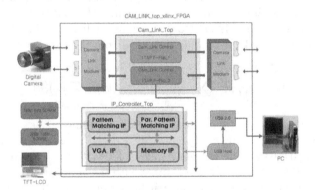

Fig. 2. Pattern Matching System Configuration

4   Experiments and Results

4.1   Experimental Environments

In this article, Xilinx ISE 11.1i Tool was used to design the Logic IP for the Pattern Matching System. Isim were used to confirm the designed IP. The 4M60 model by DALSA supporting Medium was used as the digital camera. They were ported to the desktop using AMD 2.2Ghz, RAM 4G and Windows XP, and used in the experiments.

4.2   Camera Link IP Simulation

The standard image data transmitted from the Digital Camera needs to be verified if they can perform accurate data transmission functions in the FPGA based Pattern Matching System, which is designed to support a Camera Link Medium. Figure 3 shows the simulation result of the Line 1 zone out of the Camera Link Data Image. The standard image data transmitted from the Digital Camera needs to be verified if they can perform accurate data transmission

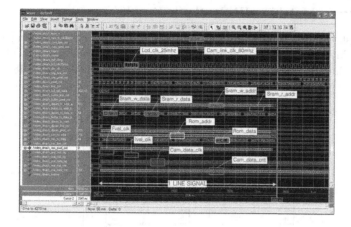

Fig. 3. Camera Link IP Simulation

functions in the FPGA based Pattern Matching System, which is designed to support a Camera Link Medium. The LVAL was counted 1,728 times from Start to End while the FVAL signal is '1'. The data was counted 588 times while the LVAL signal was '1' and printed out.

4.3   Single Clock Processing MAD Pattern Matching IP Simulation

When Single Clock Processing is not applied, the total operation times will be 3.5G. However, as the number of operations to calculate the absolute difference is reduced to 1/10, the total operation times will be the Sum of Absolute Difference ((480-64)*(480-64)*64) + Array Input (480*(480 - 64) + 64*64).

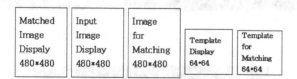

Fig. 4. Single Clock Processing MAD Pattern Matching IP Memory Configuration

5   Experimental Results

Experiments were performed to determine if Image Data transmitted in high speed from Digital Camera operate appropriately in the IP designed inside of the FPGA. In addition, it is reviewed if the data operates in the Medium method of Camera Link. Each IP presented in this article was modeled using VHDL. As all the units are designed in RTL. It is composed of a Xilinx FPGA Chip and Peripheral SDRAM, Camera Link Transmission Driver, Power Part, LED, KEY and TFT-LCD.

5.1   Pattern Matching IP Synthesis Result

Figure 5 shows the Single Clock Processing MAD Pattern Matching IP result when the MAD algorithm was applied. The template used for matching was a 64*64 size image and the target image for pattern patching was a 480*480 size

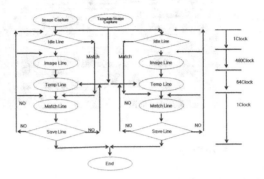

Fig. 5. Single Clock Processing MAD Pattern Matching IP Flow Chart

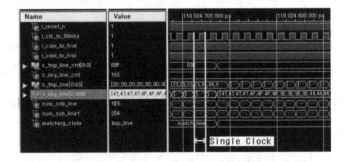

Fig. 6. Single Clock Processing MAD Pattern Matching IP Simulation

Fig. 7. Single Clock Processing MAD Pattern Matching IP Result

Table 1. MAD algorithm Comparison

	Operation Times	Operation Time (@80MHz)
MAD	(M-N)*(M-N)*N*N*C + (M*(M - N) + N*N)	44.3s
Single Clock Processing MAD Pattern Matching IP	(M-N)*(M-N)*N) + (M*(M - N) + N*N)	158ms

image. The 480*480 image at the right of the screen is the Pattern Matching Result Image and the coordinates of the matching area through the MAD were (0, 0) and (63. 63). A 64*64 image at the center of the screen is the template image. A 240*240 image at the left of the screen is the target image. In table 1, the Operation Times and Operation Time of each algorithm were compared when the target image size was M*M and the template image size was N*N. Compared to conventional MAD, the MAD Pattern Matching IP method reduces the operation times to 1/(N*C). If the Single Clock Processing MAD Pattern Matching IP is used, it decreases dramatically. If the operation is performed in 80MHz, which is the CamLink input frequency, the conventional method takes 44.3S, and Single Clock Processing MAD Pattern Matching IP takes 158ms.

6 Conclusion

A Real Time PCB Pattern Matching System using Camera Link was implemented, which is considered to be an important process in PCB manufacturing. It allows compaction, reuse and high speed processing by implementing a range of algorithms used in the visions in the FPGA module used for a Pattern Matching System as Logic. In addition, as it is designed to support FPGA-based Camera Link Medium, the image data transmitted from Camera on Windows based PC can be checked without an expensive Frame Grabber Board. The Pattern Matching method using the Single Clock Processing MAD algorithm proposed in this article allows rapid grid conformity even with moving camera image because it processes the image entered at high speed in real time. It also shows stable performance against the vibration during transportation. VGA Control IP was designed to display the image data processed by each IP on TFT-LCD, and a 460Kbyte Dual Port SRAM block was used to print out the 480*480 image. As the Real Time Pattern Matching System processes algorithms in the same way as H/W, it takes 140ms for a 480*480 image.

Acknowledgments. This research was supported by the Industry Core Technology Development Project by Ministry of Knowledge and Economy and Korea Evaluation Institute of Industrial Technology.

References

1. Jaemyung, Y., et al.: A Study on Optimal Design of Flip-chip Inosculation Equipment. Korea Society of Machine Tool Engineers (2007)
2. Seongbyeong, L., et al.: A Study on the Enhancement of Flip-chip Package Reliability. Korea Society of Machine Tool Engineers (2007)
3. Dongjung, G., et al.: Digital Image Processing using Visual C++. Sci. Tech. Media (2005)
4. Seongyu, H.: Image Processing Programming Visual C++. Hanbit Media Inc. (2007)
5. Matrox: Solios XCL Image Frame Grabber Board Manual (2006)
6. Xilinx: Virtex 4 XC4VLX200 Technical Reference Manual (2007)
7. Youngah, L.: Real-time PCB Vision Test using Pattern Matching. The Institute of Electronics Engineering of Korea Research Development Journal (2003)

A New Technique of Camera Calibration
Based on X-Target

Ruilin Bai, Jingjing Zhao, Du Li, and Wei Meng

Research Institute of Intelligent Control, Jiangnan University,
214122, Wuxi, JiangSu, China

Abstract. A new technique of camera calibration based on X-target is proposed. Calibration steps include: judging gray saturation, adjusting verticality between optical axis and object surface, two-dimensional plane camera calibration. In the calibration process, the improved Harris operator and spatial moment are used to detect sub-pixel X-target corners, and the accuracy achieves up to 0.1 pixels. It does not need to calculate the specific internal and external camera parameters, and only calculates the relations of world coordinates and image pixel coordinates, and the distortion model. The method has better accuracy and stability, and has been applied in the industrial field of embedded machine vision.

Keywords: computer vision, camera calibration, sub-pixel, optical axis verticality.

1 Introduction

Visual measurement system is critical and important part. Camera calibration technique is widely used in many fields about the machine vision, such as three-dimensional reconstruction, navigation, detection of complex three-dimensional surface, medical image analysis, and three-dimensional attitude measurement of spacecraft or missile.

The paper [1] proposed a two-step camera calibration method. Although it can calibrate camera internal and external parameters, it required the camera obtained the X-target images in different orientations. It needed nonlinear optimization, and the calculation process was complicated. The paper [2] presented a plane visual calibration based on ideal grids. Though it had small computations and high precision, it was difficult to detect accurate corners and determine precisely the ideal grid.

Based on the accuracy and stability of the two-dimensional visual measurement in specific industries, the paper proposes a highly accurate camera calibration based on single X-target image. And it acquires the measuring accuracy with 0.03mm. First it need to judge the gray saturation of the image, and then judge verticality between optical axis and object surface based on principles of pinhole imaging and digital image processing [3,4]. The improved Harris operator and spatial moment are used to detect the sub-pixel corner [5,6,7]. Finally, we do the two-dimensional visual camera calibration [2,8,9].

K. Li et al. (Eds.): LSMS / ICSEE 2010, Part I, CCIS 97, pp. 39–46, 2010.

2 Calibration Principle

2.1 Verticality Adjusting Between Optical Axis and Object Surface

Judging verticality between optical axis and object surface can decrease the deviation between optical axis and the normal line of the measured object surface, and improve the accuracy of two-dimensional vision measurement. The 3-D imaging camera model is shown in Figure 1. COC' is the camera optical axis, and AC、 FM are measured objects. C is the intersection of optical axis and measured object surface. Optical axis and plane surface 1 are vertical, and the image of the measured object AC is $A_1'C$. The θ is the angle between plane surface 1 and 2, and $A_1C=AC$.

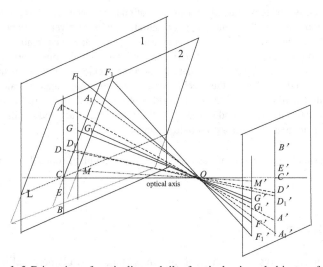

Fig. 1. 3-D imaging of verticality and tilt of optical axis and object surface

By the perspective projection principle, the image $A_1'D_1$ are given by

$$|A_1'D_1'|=|OC'|*(\frac{|AC|*\cos\theta}{|OC|-|AC|*\sin\theta}-\frac{|DC|*\cos\theta}{|OC|-|DC|*\sin\theta}) \quad (1)$$

$\theta\in[-\pi/2 \quad \pi/2]$, when the measured object tiles toward the camera lens, θ is positive. When far away from the camera lens, θ is negative.

From the formulation (1), we can have: When optical axis and object surface are vertical, the difference of the image $A'D'$ and $B_1'E_1'$ is 0 in theory. When the measured object tiles, in the region closing to the camera lens, the image $A_1'D_1'$ is longer than the image $A'D'$, and far away , the image $B_1'E_1'$ is shorter than the image $B'E'$.

We use the area method to adjust verticality. The area includes lots of information. When the lens has little tilt, area has larger changes, and affects smaller by external factors such as environmental impact and the light source. It has better stability. The rule of area variation of the image $A'D'G'F'$ is given by:

1) When optical axis and object surface are vertical, images of the same areas are equal.

2) When θ is larger and larger, the area of image $A'D'G'F'$ becomes larger.

3) When θ is smaller and smaller, the area of image $A'D'G'F'$ becomes smaller. It can judge verticality between optical axis and object surface.

2.2 Detection of Sub-pixel Corners

X-target is shown in Figure 2. The improved Harris operator is made to detect pixel-level corners. It has high stability and robustness. It only can achieve pixel-level detection. Harris operator obtains the corner response to judge the position of the corner (CRF). If CRF is greater than the threshold, and is the maximum in the local, the point is the corner. CRF is given by

$$CRF(x, y) = \det(M) - k * (trace(M))^2$$

$$M = \sum_{x,y} G(x,y) \begin{bmatrix} I_x^2, I_x I_y \\ I_x I_y, I_y^2 \end{bmatrix}, \quad G(x,y) = \exp(-\frac{x^2 + y^2}{2\sigma^2}) \tag{2}$$

The det(M) is the determinant of the real symmetric matrix M. The scale factor k is ranged from 0.04 to 0.06. I_x and I_y are gradients in the direction of X and Y.

We can use spatial-moment to get sub-pixel positions of four edges around the corner, and use line fitting to detect sub-pixel corners. A continuous two-dimensional edge specified by back-ground intensity h, edge contrast k, edge translation L, and the angle θ is shown in Figure 3. The edge is simply the step transition from gray level h to h+k. The L is defined to be the length from the center of the edge model to the step transition and is confined to the range of -1 to +1. θ specifies the angle the edge makes with respect to the x-axis. The edge is defined to lie within the unit circle.

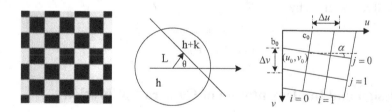

Fig. 2. X-target   **Fig. 3.** Two-dimensional ideal edge   **Fig. 4.** Ideal grid

A rotation of the circular window by θ aligns all moments containing edge information along the x-axis. This is done to reduce the dimensionality of the edge problem. We can use moment principle to get relations of original moments M_{ij} and rotated moments M'_{ij}. Original moments are estimated by correlating operation. L can be given by

$$L = \frac{4M'_{20} - M'_{00}}{3M'_{10}}, \quad \tan\theta = M_{01}/M_{10} \tag{3}$$

To consider the large rotation of the X-target, the changed template to find edge points can be used. It can use moment method to find sub-pixel edges, and then use least square method to fit four edge lines. The intersection is the sub-pixel corner.

2.3 Camera Calibration Principle

The X-target is 2D plane, so Z=0. The relationship between image pixels coordinates and camera world coordinates is given by

$$\overline{m} = [u', v'\ 1]^T, \quad \overline{M} = [x, y, 1]^T, \quad s\overline{m} = H\overline{M} \tag{4}$$

$$\begin{cases} u' = u + \delta_u \\ v' = v + \delta_v \end{cases} \tag{5}$$

The 3*3 matrix H is defined as homography matrix. (u, v) is the real point detected from image, and (u', v') is the corresponding coordinate without distortion.

Distortions include axisymmetric and non-axisymmetric[10]. Considering these distortion, we can get total distortion:

$$\begin{cases} \delta_u = k_0 \dfrac{u}{\sqrt{u^2 + v^2}} + k_1 u + k_2 u \sqrt{u^2 + v^2} + k_3 u(u^2 + v^2) + (k_4 + k_5)u^2 + k_6 uv + k_4 v^2 \\ \delta_v = k_0 \dfrac{v}{\sqrt{u^2 + v^2}} + k_1 v + k_2 v \sqrt{u^2 + v^2} + k_3 v(u^2 + v^2) + +(k_6 + k_7)v^2 + k_5 uv + k_{-7} u^2 \end{cases} \tag{6}$$

In the image center, aberrations are the smallest, and it can almost negligible. So the ideal grid field can be fitted by corners of the image center. Ideal grid is shown in Figure 4. According to ideal coordinates, we can get $k_0 \sim k_7$. The coordinate of ideal grid field can be given by

$$\begin{cases} u = -kv + b_0 + i\Delta u \\ v = ku + c_0 + j\Delta v \end{cases} \tag{7}$$

3   Design and Implementation of Camera Calibration

3.1 Design of Optical Axis Verticality

We use the center 3*3 grids of X-target. We calculate an area of the first row of the nine grid, and an area of the three row, and an area of the first column, and an area of the three column, denoted correspondingly by S_u, S_d, S_l, S_r. For the X-target image has geometrical distortion, we adopt Heron Formula to solve the special area.

Area method can not only judge whether optical axis and object surface are vertical, but also judge the relative tilted direction of camera lens and object surface. This is based on:

1) $|S_u - S_d| < threshold$ and $|S_l - S_r| < threshold$, optical axis and object surface are vertical.

2) $|S_u - S_d| > threshold$:

① $S_u > S_d$, the upside of the image closes to the camera lens, and the downside is far away.

② $S_u < S_d$, the upside of the image is far away from the camera lens, and the downside closes.

3) $|S_l - S_r| > threshold$:

①、$S_l > S_r$, the left closes to the camera lens, and the right is far away.

②、$S_l < S_r$, the left is far away from the camera lens, and the right closes.

When camera lens and measured object surface has a relatively up and down tilt, or a relatively left and right tilt, then the absolute difference of the area in the up-down region is greater than the threshold value, and meanwhile the absolute difference of the area in the left-right region is greater than the threshold value. So relative positions have the four relations such as top-left, top-right, bottom-left, bottom-right.

3.2 Design of Sub-pixel Corners Detection

Harris operator and Spatial Moment are used to detect sub-pixel corners. The process of Harris operator optimization is: In the binary image, we can find the corresponding intersection of the two borders, defined as a rough corner. To prevent inaccuracy of the rough corner location and lead to miss the corner, we should do the mask processing to the rough corner. This can compose of the operation 5*5 region of Harris operator. Table 1 reflects the different time of detecting corners in 640×480 image resolution. Harris operator after optimization can significantly reduce the computation, and improve real-time of the algorithm.

Table 1. Comparison of computing time before and after optimization

A pixel corresponding the distance (mm/pixel)	Time before optimization (ms)	Time after optimization (ms)
1/12	620	120
1/15	635	100
1/18	640	90

Optimization of detecting sub-pixel corners: judging gray saturation. The light had a large effect to the stability of corners. When the light intensity is saturated or too weak, the information of marginal parts will be lost, and it leads to corners inaccuracy. The first step needs to judge the degree of saturation of the light before detecting corners. In experiments, we can get the maximum gray need to keep between 200 to 240 pixels by observing the stability of corners, and corners are relatively stable.

The method of stored sub-pixel corners is as followed: In the camera calibration, the pixel coordinates of image corners must correspond with the actual space coordinates, so corners need to store orderly. Y direction's intercepts of all corners of the same row are the same. We can use the Y intercept to store corners by the ranks.

3.3 Design of Camera Calibration

The algorithm designs for two-dimensional measurement, so it only needs to calculate the relations of world coordinates and image pixel coordinate (H) and the distortion model. The algorithm is as followed:

1) Adjustment of light intensity: Adjust the brightness of light source, and observe the state diagram of light intensity in the real-time display interface. If the maximum of the image gray is greater than 240 pixels, the state diagram of light intensity shows in red, and it sends out the request that reduces the light intensity to down-bit machine. If the maximum is less than 200 pixels, the state diagram shows in orange. Otherwise the state diagram in green shows suitable. In the suitable light intensity, the light just needs to fine-tune.

2) Adjusting verticality between optical axis and object surface: According to the display of the operating interface, we fine-tune the lens to the verticality, and repeat adjustments to the first step and the second step to meet need.

3) Camera calibration: we use the improved Harris operator, improved spatial moment to detect sub-pixel corners, and then use corners of the image center to fit the ideal grid by principle the distortion of the image center is the smallest. At last, calculate distortion parameters and H by the Principle of two-dimensional visual calibration.

4 Experimental Results

4.1 Light Intensity and Optical Axis Verticality Results

The image resolution is 640×480, and camera lens are 8mm, 12mm, 16mm. The size of X-target is 75mm×75mm, and the size of its grid is 2mm×2mm ± 0.001mm.

The adjustment of light intensity is shown in Figure 5. Area method can not only judge whether optical axis and object surface is vertical, but also judge the relative tilted direction of camera lens and object surface. The current status of the lens is shown in Figure 6. Users can adjust the lens by the interface prompt.

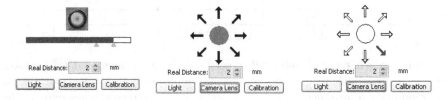

Fig. 5. Real-time image of light intensity   **Fig. 6.** Real-time image of adjusting camera lens

4.2 Calibration Results

The accuracy of corner detection has reached 0.1 pixels, and the real-time of corner detection is higher. We should calculate the maximum measurement error and standard deviation. According to the result of calibration, it measures the distance between the two grids of X- target (2mm). We calculate the standard deviation η , maximum measured value l_{max} and the average measured value $\bar{l}$. l' is the measurement distance, n=375. The accuracy measurement of X-target is shown in Figure 7. The X axis represents the number of the measurement, and the Y axis represents the error between the measuring value and the real value, and unit is millimeter.

$$l_{max} = 2.014419mm, \qquad \bar{l} = 2.0000879mm, \qquad \eta = \sqrt{\sum (l' - \bar{l})^2 / n} = 0.003545\,mm \qquad (8)$$

From the Figure 7, the measurement errors are ranged form -0.005 to 0.005, only few errors are above 0.01mm.

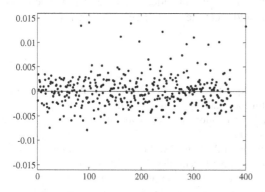
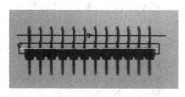

Fig. 7. Measurement accuracy of X-target   **Fig. 8.** Measurement of pins

Different specifications of pins (2.54mm) are measured. It is shown in Figure 8. In embedded machine vision systems, click on the line position of the platform, to find the edge lines as the base line, and do measurement between the lines. Measurement data records in the following Table 2.

Errors in Table 2 are measured when the ratio of the actual distance and image distance is 1:10. This method does not require nonlinear optimization, and it can reduce the computational complexity. The measuring accuracy can reach 0.03mm. It can meet the need of general industrial field.

Table 2. Measurement data of pins

	1	2	3	4	5
Distance(mm)	2.5667	2.5274	2.5631	2.5681	2.5622
error(mm)	0.0267	0.0126	0.0231	0.0281	0.0222
	6	7	8	9	10
Distance(mm)	2.5637	2.5165	2.5158	2.5619	2.5628
error(mm)	0.0237	0.0235	0.0242	0.0219	0.0228

5   Conclusions

1) Using the area method to judge verticality of optical axis and object surface. It can not only judge whether optical axis and object surface are vertical, but also accurately judge the relative tilted direction of camera lens and object surface.

2) Combining the improved Harris operator and spatial moment to detect sub-pixel corners. It has higher accuracy, stability and robustness, and it reduces the relatively computation. The accuracy of corner detection has reached 0.1 pixels.

3) Judging gray saturation. This is an optimization of corner detection to prevent that the light intensity is too strong (weak), leading to lose the information of the marginal part, making the corner instability. It can better improve corner stability, improve calibration accuracy.

4) In paper, two-dimensional visual calibration has higher accuracy, simple operation, good stability, less computation, and has no other ancillary facilities. This algorithm has been applied in the industrial field of embedded machine vision. The measurement accuracy can reach 0.03mm, and meet the need of general industrial field.

References

1. Zhang, Z.Y.: A Flexible New Technique for Camera Calibration. IEEE Transaction on Pattern Analysis and Machine Intelligence 22(11), 1330–1334 (2002)
2. Quan, T.H., Yu, Q.F.: High-accuracy Calibration and Correction of Camera System. Acta Automatica Sinica 26(6), 748–755 (2000)
3. Ryoo, J.R., Doh, T.Y.: Auto-adjustment of the Objective Lens Neutral Position in Optical Disc Drives. IEEE Transactions on Consumer Electronics, 1463–1468 (2007)
4. Murata, N., Nosato, H., Furuya, T., Murakawa, M.: An automatic multi-objective adjustment system for optical axes using genetic algorithms. In: Proceedings of the 5th International Conference on Intelligent Systems Design and Applications, pp. 546–551 (2005)
5. Harris, C., Stephens, M.: A Combined Corner and Edge Detector. In: Proceedings of the 4th Alvey Vision Conference, Manchester, pp. 147–151 (1988)
6. Lyvers, E.P., Mitchell, O.R.: Subpixel Measurements Using a Moment-Based Edge Operator. IEEE Transactions on Pattern Analysis and Machine Intelligence (0162-8828) 11(12), 1293–1309 (1989)
7. Chen, D., Wang, Y.: A New Sub-Pixel Detector for Grid Target Points in Camera Calibration. In: Conference on Optical Information Processing, China, vol. 6027 (2006)
8. Zhang, H.T., Duan, F.J.: Vision Model Calibration Based on Grid Target. Opto-Electronic Engineering 33(11), 57–60 (2006)
9. Nie, K., Liu, W., Wang, J.: Camera Calibration Method with a Coplanar Target and Three-dimensional Reconstruction. In: International Conference on Optical Instruments and Technology: Optoelectronic Imaging and Process Technology, China, vol. 7513 (2009)
10. Weng, J.Y., Paul, C., Marc, H.: Camera Calibration with Distortion Models and Accuracy Evaluation. IEEE Transactions on Pattern Analysis and Machine Intelligence 14(10), 965–980 (1992)

Application Research of the Wavelet Analysis in Ship Pipeline Leakage Detecting

Zhongbo Peng[*], Xin Xie, Xuefeng Han, Xiaobiao Fan

College of Maritime, Chongqing Jiaotong Univercity, Chongqing China
Pzbpzbpzb751123@163.com

Abstract. Monitoring of ship pipeline leakage detecting is one of the most important techniques to be developed as it can help to prevent damages of ship working safe. Negative pressure wave technique is an effective method for paroxysmal fluid leakage detection and location. However, it is difficult to distinguish sources which led to the fluid pressure drop. In order to solve the problem, wavelet transform algorithm was adopted to define inflexion of the negative pressure wave when it propagates along the pipe, and wavelet threshold denoise technique was used to separate the characteristic inflexion of negative pressure wave when calculating the leaking position. A new pipeline detection and location system on the basis of that was developed.

Keywords: Negative pressure wave; Wavelet Transform.

1 Introduction

The piping of ship is an important part, like blood vessels for the body, which is the main contact and auxiliary equipment. To make a good run of these devices, they replenish the necessary fuel, lubricants, cooling water and compressed air and so on constantly.They are divided by a variety of systems. The large number of pipes locate in various parts of the ship, with the characteristics of small space and the complication of maintenance.Sailing on the water in the whole year, piping is in the environment of humid, high temperature, vibration and harsh. It is difficult to avoid leakage, so piping is the outstanding problems about the safety of ships. The slight effect is ranging from equipment leakage, serious result is equipment damage or structural failure occurred even sank.

At present, there are not yet effective measures to avoid piping leakage. Usually, observing a variety of piping and maintenance of pipe line on a regular basis by the crew can not accurately determine the specific circumstances of leakage. They are carried out after the failure to deal with leakage. States have invested a great deal of manpower and funding for piping corrosion and protection, mainly in the research of chemical piping corrosion. At present there are not yet technology and equipment for real-time online monitoring system and fault diagnosis in ship pipeline.

[*] The research is supported by Natural Science Foundation of Chongqing (CSTC2007BB2428) and Foundation of Chongqing Educational Commission(KJ00402).

K. Li et al. (Eds.): LSMS / ICSEE 2010, Part I, CCIS 97, pp. 47–55, 2010.
© Springer-Verlag Berlin Heidelberg 2010

Leakage monitoring techniques have been the focus of researchers throughout the world along with the construction of pipelines since leakage of oil pipeline is a crucial problem in the oil and chemical industry. Because it may result in great economic loss as well as environmental pollution, there is much interest in detecting the leakage rapidly and locating the leak point precisely. At the present time, some physical methods such as baric gradient method, negative pressure wave method, flow balancing method and ultrasonic guided wave method, as well as some chemical methods, have been successfully applied in pipeline leakage monitoring [1]. These leakage monitoring methods differ from each others incharacters and application occasions.

Negative pressure wave technique is an effective tool for pipeline leakage detection and location [2]. However, the monitoring process may be interfered by either noise or some normal operations. In this paper, a monitoring and positioning system based on wavelet analysis is developed for ship pipeline.

2 The Principle of Leak Detection

When the pipes leak, the pipeline pressure of the first side has decreased, while flow increases. The pressure of the end decreases, flow will also decrease. These are different from the corresponding pressure and flow changes when pumps and valves are adjusted. When the rate of pump increases, pressure of the first side and flow increase. Also the pressure of the end and flow increase. When the rate of pump reduces, pressure of the first side and flow rate drop. Also the pressure of the end and flow drop. Therefore, this method can effectively avoid false alarm by pipeline open and close.

In this article, the measure of flow uses the ultrasonic flowmeter of time difference clamp-type. This flowmeter is accuracy, simple structure and can be easily installed in the pipeline outside for non-contact measurement without any disturbance. So it is easy to meet users' requirements. When the pipes leak, there is difference pressure at the leak point. As a result, the fluid loss quickly destroying the balance of the original hydraulic system, so the pressure drop and there is transient fluctuations in the flow at both ends of the pipe. Due to difference pressure, the liquid closing to leakage points add up the process followed by the downstream transmission. In other words, a negative pressure wave transmission is produced at the leakage point. The method of transient flow detection bases on the time difference of the transient pressure generated by the wave from upstream to downstream and the speed of pressure wave in the pipe to calculate the location of the leakage point.

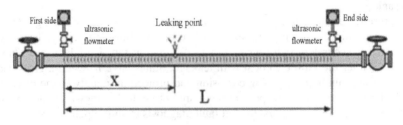

Fig. 1. The principle of negative pressure wave positioning

As long as it can accurately capture the information about leakage by both ends of transient flow, the leakage can be monitored and located according to the speed of time and pressure wave. The method is very fast for responsing speed and high positioning accuracy.

The positioning principle is shown in Fig.1, and the leaking point in the pipeline can be calculated as follows,

$$X = \frac{1}{2}(L + V\Delta t) \tag{1}$$

in which, X is the distance from the leaking point to sensor1 installed in the former side. L is inspection length of the pipeline(here, it is the distance between two sensors), and V is propagation velocity of the pressure waves in pipe mass media, Δt is the time delay between the reception of pressure waves by sensor in up and down streams.

However, there are a lot of shortages in monitoring of flow. Firstly, with the change of the parameters of the external environment, flux signals changes in the larger random interference. Also, it is less sensitive. Secondly, at this stage, the most methods of flux monitoring introduce decision of threshold-function. This function collects the real-time flux signals comparing to the threshold. When the threshold is not in the threshold limit, it will handle accordingly. The principle of the method is simple, real-time, accurate and reliable. But, the threshold should change with the specific situation for the different pipelines or media. It is not a single static value. That determines the threshold is uncertain in the function. Furthermore, the threshold is determined after a certain number of experiments. These are not economic and impossible for variety circumstances.

With the characteristics of time and frequency domain at the same time and self-adjusting the width of window, wavelet transform can focus on any details. It is good adaptability [2]. To achieve the purpose of real-time monitoring, we must get the real-time wavelet algorithm of wavelet coefficients using improved real-time algorithm in the actual monitoring. At the same time, because of the uncertainty of threshold-function in of decision-making, we test the wavelet coefficients using the related real-time detection. Then we measure the leakage of pipes through the characterization of a number of mutual relations between the degree of correlation in mathematical statistics.

3 Wavelet Algrithm Applied in the Monitoring System

3.1 Mallat Algorithm and Its Application

Mallat algorithm [3],[4], also known as pyramidal algorithm, which consists of wavelet filters H, G and h, g ; decompose and reconstruct the signal, the specific decomposition algorithm is as in (2).

$$\begin{cases} A_0[f(t)] = f(t) \\ A_j[f(t)] = \sum_k H(2t-k)A_{j-1}[f(t)] \\ D_j[f(t)] = \sum_k G(2t-k)A_{j-1}[f(t)] \end{cases} \tag{2}$$

In the formula, t for the discrete-time serial number,t=1,2,...,N, f(t) as the original signal, in this system, f(t) is the real pressure waves of the signal; j as the number of layers, j=1,2,...,J, J=log2N, but the system j value of 4, through the test, j= 4 is found to be well positioned to meet testing needs, the greater j is, the higher the testing accuracy is, but calculation overhead of system is more so that not meet the real-time. while j is too small and can not achieve a normal diagnosis. H, G is time-domain wavelet decomposition filter, is actually filter coefficients; Aj is wavelet coefficients of the j-layer approximation part (ie, low-frequency part). Dj is wavelet coefficients of details part of the layer (ie, high-frequency part).this system uses the DB4 as the wavelet coefficients. The meaning of the formula [5] is: Assuming the detection of discrete signals f(t), which the actually pressure waves values is A0, approximation components of signal f(t) in section 2j (j-layer), namely wavelet coefficients Aj of low-frequency part, it is achieved by approximate part wavelet coefficients of 2j-1(the j-1-layer) convolution with decomposition filter H, and then it is obtained by every other point sampling convolution results. While approximate part wavelet coefficients of 2j-1(the j-1-layer) convolution with decomposition filter G, the signal Dj is obtained by every other point sampling convolution results. in the scale 2j (or j-layer), the signal f(t) is decomposed into approximation part of the wavelet coefficients Aj (in the low-frequency sub-belt) and the details of the part of the wavelet coefficients Dj (high-frequency sub-belt). Decomposition algorithm [6] can be expressed in Fig. 2.

Mallat algorithm has the advantage of a direct calculation.

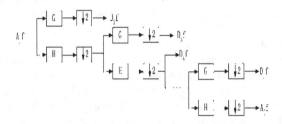

Fig. 2. Mallat decomposition algorithm

In the monitoring process, to record full normal pressure waves information curve of the processing is needed. Then, use Mallat algorithm to the curve with wavelet analysis, and use pressure waves information of the wavelet transform results as standard of monitoring determination.

3.2 Precise Acquisition of the Time Difference of Pressure Waves Propagating to Up and Down Streams

For precise acquisition of time difference between pressure waves acquired by transducers installed up and down stream, it is required to precisely capture the corresponding characteristic points in the signal sequence of negative pressure waves caused by leakage. Because of inevitable electromagnetic interference at ship site and vibration of oil transfer bump, a great deal of noise is attached to the signal sequence of collected pressure waves, and how to precisely pick up characteristic points of the signals from noise is the key to positioning [6].

Wavelet Transform is used to extract edge of instantaneous pressure wave. By extracting characteristic point in measuring signals captured at only one end, satisfying result can be achieved. The continuous wavelet transform of x(t) is

$$WT_a x(t) = \frac{1}{a} \int x(\tau) \phi(\frac{t-\tau}{a}) d\tau \qquad (3)$$

It can also be written as

$$WT_a x(t) = x(t) * \phi_a(t) \qquad (4)$$

In which, $\phi_a(t) = \frac{1}{a}\phi(\frac{t}{a})$ is the flex of basic wavelet transform in scale a. The aim of monitoring is to detect the singularity with high sensitivity. The first derivative of Gauss low-pass function $(\theta(t))$ is selected as basic wavelet transform function.

$$\theta(t) = \frac{1}{\sqrt{2\pi}} e^{-t^2/2} \qquad (5)$$

$$\phi(t) = \frac{d\theta(t)}{dt} = \frac{1}{\sqrt{2\pi}} t e^{-t^2/2} \qquad (6)$$

It can be demonstrated that:

$$WT_a x(t) = x(t) * \frac{d\theta_a(t)}{dt} = a\frac{d}{dt}[x(t) * \theta_a(t)] \qquad (7)$$

The continuous wavelet transform of x(t) equals to the following process. First, x(t) is filtered by a low-pass filter, and then it is differentiated. The extreme point of WTax(t) corresponds to the inflection of smoothed signals. The signals measured on site are usually the superposition of deterministic signals and stationary noises. The extreme value of wavelet transform of noises goes to zero with increase of the scale, however, the extreme value of useful signals will increase or hold the line with scale's increase. Even when extreme value of useful signals attenuates together with noises, the former's attenuation rate will be slower than noises. Therefore, the changed edge of signals which corresponds to extreme value of wavelet transform could be found in large scale.

3.3 Distinguish of Pressure Drop Causing by Valve Regulation and Leakage

Successfully positioning the leaking point is crucial to the leakage detection and location system. As we all have known, it is a complex environment in the pipeline site and the monitoring process may be interfered by either noise or some normal pipeline management operations, such as pump adjustment and valve adjustment. And the key to positioning is picking up characteristic points of the signals from noise or normal operations. It is demonstrated that Wavelet Transform is an effective way to detect instantaneous negative pressure wave in low energy. The quick speed algorithm of wavelet transform (Mallet Algorithm) is used for multi-resolution analysis of negative pressure wave, and low-frequency general picture and highfrequency part of the collected signals can be obtained.

Locating the negative pressure wave is to define the position of minimum value among the high-frequency signals. The diagrams of pressure drop signals caused by valveregulation and leakage are shown in Fig.3 (a) and (b). In the time-domain diagram, it is difficult to distinguish the valve regulation signal from leakage signal. However, when the two signals were decomposed in three-level and reconstructed using wavelet package algorithm (see fig. 4), eigenvectors in different frequency-band of the signals can be extracted. By establish coincidence relation between eigenvectors and different source signals, sources that causing pressure drop can be distinguished. And this could be used as an effective judgment to distinguish the two signals.

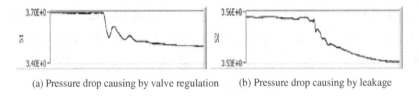

(a) Pressure drop causing by valve regulation   (b) Pressure drop causing by leakage

Fig. 3. Diagram of pressure drop signals

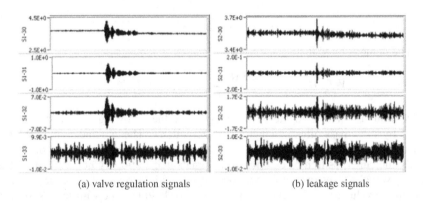

(a) valve regulation signals   (b) leakage signals

Fig. 4. Level wavelet package decomposition and reconstruction signals

4   The Use of Cross-Correlation Detection in Monitoring

Most of the real-time monitoring systems compare numerical data to the standard real-time monitoring data using the method of decision-making system in order to obtain real-time status. However, in pipeline monitoring systems, these methods aren't the versatility. In the signal processing, the correlation coefficient can be used for the characterization of signals similar. Experiments show that this way is good.

The description of cross-correlation function is the relation of two different stochastic processes at different times, while the number of mutual relations shows the close of two signals. In mathematical statistics, we use the standard shown in table 1.

Table 1. The range of the related coefficient vs. the degree of relation

The range of the Related coefficient	The degree of relation
1.00	Entirely related
0.70-0.99	Highly related
0.40-0.69	Moderately related
0.10-0.39	Low related
Less than 0.10	Weak or no related

At upstream and downstream, we set sensors apart, named A, B. We get the signal at the first side in the pipeline (x-point) and the end side (y-point), then name the sample-based function x(n) and y(n), means two sides of the wavelet coefficients [7].

Their mutual relation is defined as follow:

$$\rho = \sum_{n=1}^{N} x(n)y(n) / \sqrt{\sum_{n=1}^{N} x(n)^2 \sum_{n=1}^{N} y(n)^2} \tag{8}$$

When there is no leakage, ρ will be value greatly. When the leak occurred, ρ will be small.

5 Identification of a Small Pipeline Leak

When pipeline leaks, the flux signal shows a certain degree of volatility. In order to achieve real-time monitoring, we only need to pay attention to the abnormal increase or decrease about information flows. So the algorithm needs to be sensitive for abnormal changes.

If information flows decline abnormal, the signal that the sensor collected in the end side about the low-frequency parts of A4 and the high-frequency parts of D4 are unusual, as Figure 5.

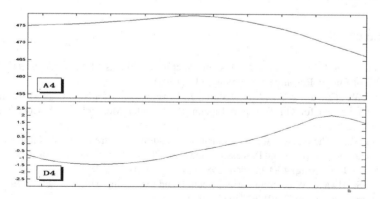

Fig. 5. The wavelet coefficient in the end side A4 and D4 in pipeline leakage

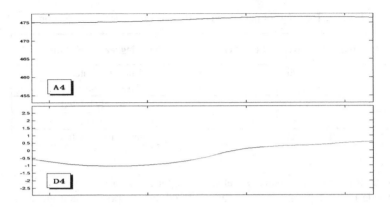

Fig. 6. The wavelet coefficient in the first side A4 and D4 in pipeline leakage

The collection of wavelet coefficients about A4 and D4 from the first side are shown in Fig. 6.

In Fig. 5 and Fig. 6, parts of the A4 calculated by the mutual relations, the relative coefficient is less than 0.5, and D4 is also. It can be concluded a pipeline leak occurred. As usual, the two correlation coefficient about A4 and D4 reach more than 0.9.

6 Conclusion

In real-time monitoring of pipeline leakage, the collection of flux signals and the analysis of real-time wavelet can be achieved for a ship to monitor pipeline leakage, also wavelet analysis could be used as an effective judgment to distinguish of pressure drop causing by valve regulation and leakage. The use of wavelet analysis and the sensor data of the two sides about cross-correlation detection, appears to avoid the characteristics of delayed response for the flow of time domain signal. Also, it does not require tedious experiments. So it is good versatility.

References

1. Kiuchi, T.: A Leak Location Method of Pipeline by Means of Fluid Transient Model. Journal of Energy Resource Technology, 115 (1993)
2. Li Kun, W., Cuiyun, Z.Y.J., Li, Z., Shijiu, J.: Development of a New Leak Monitoring & Location System for Oil Transport Pipelines. Computer Measurement & Control 10(3), 152–155 (2002)
3. Shensa, M.J.: The discrete wavelet transform: wedding the a trous and Mallat algorithms. IEEE Transactions on Signal Processing (see also IEEE Transactions on Acoustics, Speech, and Signal Processing) 40(10), 2464–2482 (1992), doi:10.1109/78.157290
4. Mallat, S., Hwang, W.L.: Singularity detection and processing with wavelets. IEEE Trans. on Information Theory 38(2), 617–643 (1992)

5. Mallat, S., Zhong, S.: Character of signal frommultiscale edbes. IEEE Trans. on Pattern Anal. Mach. on Intell. 14(7), 710–732 (1992)
6. Hallatschek, K., Zilker, M.: Real time data acquisition with transputers and PowerPCs using the wavelet transform for event detection. IEEE Transactions on Nuclear Science, Part 1 45(4), 1872–1876 (1998)
7. Hao, X., Bai, Y., Cui, Z.: Signal processing method of correlation detection for ranging system based on combined modulation. In: 7th World Congress on Intelligent Control and Automation, WCICA 2008, June 25-27, pp. 1373–1377 (2008)

Analysis and Implementation of FULMS Algorithm Based Active Vibration Control System*

Zhiyuan Gao, Xiaojin Zhu**, Quanzhen Huang, Enyu Jiang, and Miao Zhao

School of Mechatronics Engineering and Automation, Shanghai University
Shanghai, 200072, P.R. China
mgzhuxj@shu.edu.cn

Abstract. Considering the passive vibration control methods are not effective for low frequencies and will increase the size and weight of the system, an active vibration control (AVC) system is designed in this paper based on the filtered-u least mean square (FULMS) algorithm. Giving the multi-in multi-out (MIMO) FULMS controller structure and taking the configured smart beam with surface bonded lead-zirconate-titanate (PZT) patches as research object, an AVC experimental platform is established to testify the effectiveness of the proposed controller. Experimental results indicate that the designed MIMO FULMS vibration controller has a good control performance to suppress the vibration significantly with rapid convergence.

Keywords: active vibration control, flexible structures, FULMS algorithm, piezoelectric actuators.

1 Introduction

To reduce the costs inherent with transport in the space, lighter structures are urgently needed for the future spacecraft, aerospace and aircraft components. Along with the weight reducing, it is unavoidable that the rigidity also decreases, and the vibration problem is more easily caused by external disturbances [1-4]. Thus the elimination of unwanted vibrations becomes an important issue. Though passive and active vibration control methods consist of the two main groups of structural vibration control methods, active vibration control methods are more attractive for the vibration control application of the space flexible structures, considering the passive vibration control methods are not effective for low frequencies, and many of these methods will increase the size and weight of the system [5].

The organization of this paper is as follows. In section 2, the basic form of the FULMS algorithm is introduced, and with the MIMO FULMS algorithm analyzed, the multi-in multi-output controller structure is presented. In section 3, the actual

---

* This research is supported by program of National Nature Science Foundation of China (No.90716027).
** Corresponding author.

K. Li et al. (Eds.): LSMS / ICSEE 2010, Part I, CCIS 97, pp. 56–63, 2010.

MIMO control FULMS experiment is carried out to suppress the vibration of the cantilever smart beam with the reference signal extracted from the transducer boned on the smart beam surface. Section 4 presents the conclusions.

2 MIMO FULMS Control Algorithm

In order to suppress the structural vibration, active vibration control system using adaptive filter structure outputs the control signal to the actuators according to the reference error signal. The schematic diagram of the adaptive vibration control system is shown in Fig. 1. $P(z)$ represents the primary path, $H(z)$ represents the control path, while $x(k)$ represents the reference signal.

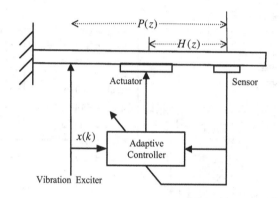

Fig. 1. Schematic diagram of the adaptive AVC system

FXLMS control algorithm has drawn wide attention in the active vibration and noise control field [6-8], with its high control correction rate, quick non-stationary response and the ability of fast tracking structural parameters and external distur- bance, But when feedback is present, FULMS control algorithm is more suitable as IIR filters require fewer arithmetic operations by making it possible to obtain well- matched characteristics with a lower order structure [9]. The MIMO FULMS algorithm can be derived as

$$E(k) = B(k) - W(k)R^T(k) - D(k)G^T(k) \tag{1}$$

$$Y(k) = \tilde{X}(k) + \tilde{Y}(k) \tag{2}$$

$$W(k+1) = W(k) + 2\mu E^T(k)R^T(k) \tag{3}$$

$$D(k+1) = D(k) + 2\alpha E^T(k)G^T(k) \tag{4}$$

Here, $E(k)$ is the L-th order residual error sequence at time k. $Y(k)$ is M-th order output sequence. $B(k)$ is the structural vibration response without control at time k. $X(k)$ filtered by W is denoted as $\tilde{X}(k)$, $Y(k)$ filtered by D is denoted as $\tilde{Y}(k)$. W and D are the FULMS controller weighting coefficient vector corresponding to the IIR filter response sequence coefficients. While the length of the two filters W and D are P and Q respectively. $\hat{H}$ is the identification model of the L-input M-output control channel. The block diagram of FULMS controller with L sensors and M actuator groups is shown in Fig. 2.

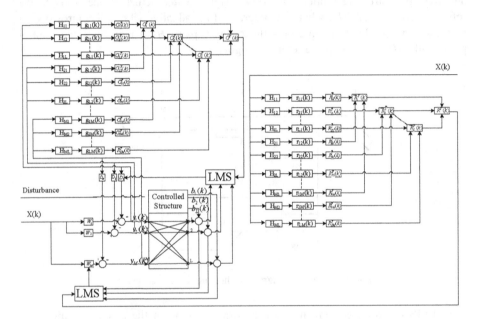

Fig. 2. Block diagram of MIMO FULMS algorithm

While

$$W(k)=\begin{bmatrix} w_{10}(k) & w_{11}(k) & \cdots & w_{1(P-1)}(k) \\ w_{20}(k) & w_{21} & \cdots & w_{2(P-1)}(k) \\ \cdots & \cdots & \ddots & \cdots \\ w_{M0}(k) & w_{M1}(k) & \cdots & w_{M(P-1)}(k) \end{bmatrix} \quad D(k)=\begin{bmatrix} d_{11}(k) & d_{12}(k) & \cdots & d_{1Q}(k) \\ d_{21}(k) & d_{22}(k) & \cdots & d_{1Q}(k) \\ \cdots & \cdots & \ddots & \cdots \\ d_{M1}(k) & d_{M2}(k) & \cdots & d_{MQ}(k) \end{bmatrix} \quad (5)$$

$$R(k)=\begin{bmatrix} R_{11}(k) & R_{12}(k) & \cdots & R_{1L}(k) \\ R_{21}(k) & R_{22}(k) & \cdots & R_{2L}(k) \\ \cdots & \cdots & \ddots & \cdots \\ R_{M1}(k) & R_{M2}(k) & \cdots & R_{ML}(k) \end{bmatrix} \quad G(k)=\begin{bmatrix} G_{11}(k) & G_{12}(k) & \cdots & G_{1L}(k) \\ G_{21}(k) & R_{22}(k) & \cdots & G_{2L}(k) \\ \cdots & \cdots & \ddots & \cdots \\ G_{M1}(k) & G_{M2}(k) & \cdots & G_{ML}(k) \end{bmatrix} \quad (6)$$

$$R_{ml}(k) = [r_{ml}(k), r_{ml}(k-1), \cdots\cdots, r_{ml}(k-P+1)] \tag{7}$$

$$r_{ml}(k-p+1) = \sum_{i=0}^{P-1} h_{mli} x(k-i-p+2) , m=1,2,\cdots M ; l=1,2,\cdots L \tag{8}$$

$$G_{ml}(k) = [g_{ml}(k), g_{ml}(k-1), \cdots\cdots, g_{ml}(k-Q+1)] \tag{9}$$

$$g_{ml}(k-q+1) = \sum_{i=1}^{Q} h_{mli} y(k-i-q+1) , m=1,2,\cdots M ; l=1,2,\cdots L \tag{10}$$

According to least mean square algorithm, the weight coefficients update equation of MIMO FULMS algorithm is shown as equation (3) and (4). μ and α are step convergence factors The convergence factor μ and α is vital for the convergence of the control algorithm. The higher μ and α is, the higher the convergence rate would be. But if μ and α is too big, the control system may not converge. Since the error surface for $E\{e^2(n)\}$ of the FULMS algorithm is multimodal, the algorithm may converge to a local minimum, which need further study or improving.

3 MIMO FULMS Vibration Control Experiment

3.1 Construction of the Experiment Platform

In our research a set of devices and apparatus are employed to develop the vibration suppression experiment platform.

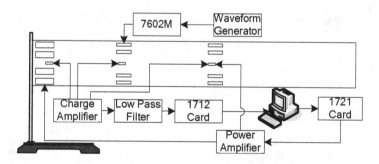

Fig. 3. Schematic diagram of the experiment setup

To guarantee satisfaction of real-time requirements for active vibration control, a personal computer is employed with the plug-in Advantech PCI-1712 and PCI-1721 card. The PCI-1712 card is a high-speed multifunction DAS card with a 1MHz 12-bit A/D converter. The PCI-1721 card is a high-speed analog output card equipped with a 12-bit, double-buffered DAC. The PCI-1712 card is connected by a shielded cable

PCL-10168 and a 68-pin SCSI-II wiring terminal for DIN-rail mounting ADAM-3968, as well as the PCI-1721 card. The schematic diagram of the experiment setup is shown in Fig. 3.

The Yong's modulus of the epoxy beam is 22Gpa, Poisson's ratio is 0.3, the density is 2100 kg/m³, and the dimension of the epoxy beam is 950mm*120mm*2mm. Considering that the vibration energy primarily concentrates on the low frequency modal, only the first four bending modes are considered in this paper. The first piezoelectric sensor is placed 42mm off the root, the second actuators group is placed at the position 235mm from the left edge, the third group is at 485mm from the left edge. To verify the effectiveness of the 2-input 2-output FULMS control algorithm, the first and the third group are chosen as the vibration control actuators, the second group is chosen as the exciter exerting disturbance.

The programming software platform used in our research is Visual C++ 6.0, with which the control algorithm is programmed and management of the PCI-1712 AD card and PCI-1721 DA card is achieved.

3.2 MIMO FULMS Simulation

It is almost impossible in practical applications to extract the reference signal from the exciter or the reference transducer that does not suffer the feedback influence from the actuators. Thus a controller would have a wide application field, if it could extract the reference signal from the transducers boned or embedded in the structures. To illustrate the mechanism why the FULMS algorithm could extract the reference from the vibration response, simulation of FULMS algorithm is done and the simulation result is analyzed. Also for convenience only one control channel is analyzed below.

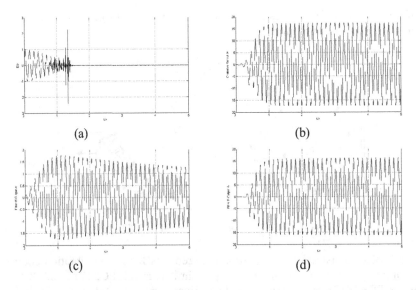

(a)   (b)

(c)   (d)

Fig. 4. FULMS simulation result when the reference signal is extracted directly from the vibration response: (a) Control performance of FULMS algorithm; (b) FULMS algorithm controller output; (c) Filter W output; (d) Filter D output

As shown in Fig. 4 (a), when the reference signal is extracted from the sensors boned on the beam surface, the FULMS algorithm is effective for the vibration control. FULMS exhibits fast convergence. As the Figure 4 (c) and (d) show the output of filter W is reflected by the actuators output, but filter D output the desired output and play a main role in the controller output. That means though the filter W is invalid, the filter D is still effective while the reference signal is extracted directly from the vibration response of the smart beam

3.3 MIMO FULMS Vibration Control Experiment

To verify the effectiveness of the MIMO FULMS algorithm based control system, the control experiment is done on the experimental platform introduced in section 3.1. The sampling frequency of the control system is 1000Hz. The second group piezo-electric actuators are used to generate the continuous external disturbance below 100Hz.

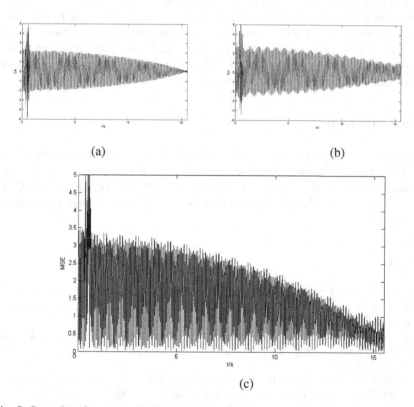

(a)   (b)

(c)

Fig. 5. Control performance of MIMO FULMS algorithm while the reference signal extracted from the reference transducer boned on the smart cantilever beam: (a) Control performance of channel one; (b) Control performance of channel two; (b) Overall control performance

Fig. 5 (a) and (b) shows the control performance of channel one and channel two. The algorithm could converge in 15 seconds, while the reference signal is extracted from the structural vibration response. Figure 5 (c) shows the weighted mean square error (the overall vibration response of the smart cantilever beam). We can see that, vibration of the smart beam is suppressed to a great extent.

In our research, the MIMO FULMS algorithm based controller shows good control performance and rapid convergence rate in most cases. But the convergence of FULMS is bad under some conditions, and the instability may occur. Thus, the further study to improve the FULMS algorithm is still needed.

4 Conclusions

This paper presents the analysis of the FULMS algorithm comparing with the FXLMS algorithm, and gives the implementation of the MIMO vibration controller based on FULMS algorithm. Both the simulation and the actual experiment results show that the MIMO FULMS algorithm could suppress the vibration of the smart cantilever beam by a bonded PZT transducer extracting the reference signal. Though the research in this correspondence is a good attempt for the implementation of the active vibration control systems, the FULMS algorithm may converge to a local minimum, and need further study and improving.

Acknowledgments

This research is supported by National Nature Science Foundation of China (No.90716027), Mechatronics Engineering Innovation Group project from Shanghai Education Commission, Shanghai University "11th Five-Year Plan" 211 Construction Project and Shanghai Key Laboratory of Power Station Automation Technology.

References

1. Wang, C.G., Mao, L.N., Du, X.W., He, X.D.: Influence Parameter Analysis and Wrinkling Control of Space Membrane Structures. Mechanics of Advanced Materials and Structures 17, 49–59 (2010)
2. Zemlyakov, S.D., Rutkovskii, V.Y., Sukhanov, V.M.: Some questions of control of the robotized in-orbit assembly of large space structures. Automation and Remote Control 67, 1215–1227 (2006)
3. Di Gennaro, S.: Output stabilization of flexible spacecraft with active vibration suppression. IEEE Transactions on Aerospace and Electronic Systems 39, 747–759 (2003)
4. Lee Glauser, G.J., Ahmadi, G., Layton, J.B.: Satellite active and passive vibration control during liftoff. Journal of Spacecraft and Rockets 33, 428–432 (1996)
5. Xia, Y.A.: Ghasempoor: Adaptive active vibration suppression of flexible beam structures. Proceedings of the Institution of Mechanical Engineers Part C-Journal of Mechanical Engineering Science 222, 357–364 (2008)

6. Qiu, X.J., Hansen, C.H.: A study of time-domain FXLMS algorithms with control output constraint. Journal of the Acoustical Society of America 109, 2815–2823 (2001)
7. Gupta, A., Yandamuri, S., Kuo, S.M.: Active vibration control of a structure by implementing filtered-X LMS algorithm. Noise Control Engineering Journal 54, 396–405 (2006)
8. Xiao, Y.G., Ma, L.Y., Hasegawa, K.: Properties of FXLMS-Based Narrowband Active Noise Control With Online Secondary-Path Modeling. IEEE Transactions on Signal Processing 57, 2931–2949 (2009)
9. Sun, X.G.M.: Steiglitz-Mcbride type adaptive IIR algorithm for active noise control. Journal of Sound and Vibration 273, 441–450 (2004)

Performance Analysis of Industrial Wireless Networks Based on IEEE 802.15.4a

Tongtao Li[1], Minrui Fei[1,*], and Huosheng Hu[2]

[1] Shanghai Key Laboratory of Power Station Automation Technology,
School of Mechatronics Engineering and Automation,
Shanghai University, Shanghai, China 200072
[2] School of Computer Science & Electronic Engineering
University of Essex, Colchester CO4 3SQ, United Kingdom
Tongtao.li@gmail.com

Abstract. The IEEE 802.15.4a standard provides a framework for low data rate communication systems, typically sensor networks. In this paper, we have established a realistic environment for the preliminary performance analysis of the IEEE 802.15.4a. Several sets of practical experiments are conducted to study its various features, including the effects of 1) numeral wireless nodes, 2) numeral data packets, 3) data transmissions with different upper-layer protocol. Time-delay is investigated as the most important performance metric. The results show that IEEE 802.15.4a is suitable for some industrial applications which have more relaxed throughput requirements and time-delay.

Keywords: performance analysis, IEEE 802.15.4a, industrial wireless network.

1 Introduction

Wireless communication has been pervading many application areas for the past few years and is affecting an ever-increasing number of aspects of daily life. New products and services concerning mobile communication (i.e., mobile audio, video, and data exchange services and the relevant devices) appear on the market almost every day, and people are getting used to relying on wireless technology in their business and entertainment activities [1]. In general, the wireless network has followed a similar trend due to the increasing exchange of data in services such as the Internet, e-mail, and data file transfer. The capabilities needed to deliver such services are characterized by an increasing need for data throughput. However, other applications in fields such as industrial [2], vehicular, and residential sensors [3] have more relaxed throughput requirements. Moreover, these applications require lower power consumption, low complexity wireless links for a low cost (relative to the device cost). IEEE 802.15.4a [4] is the one that satisfies these types of requirements.

---

* Corresponding author.

K. Li et al. (Eds.): LSMS / ICSEE 2010, Part I, CCIS 97, pp. 64–69, 2010.

The 802.15.4a physical layer is based on two different technologies: Ultra Wide Band (UWB) and chirp signals[5,6]. A standard device will be capable of transmitting in at least one of three 500 MHz-wide bands centered at 499.2 MHz, 4.493 GHz, and 7.987 GHz. Robust and energy-detection receivers for IR(impulse radio)-UWB transmission have been presented and analyzed in [7,8,9,10,11]. Taking the impact of the characteristics of the the new physical layer on medium access into account, Luca De Nardis et al. provided an overview and comparison of 802.15.4 with 802.15.4a on the MAC layer in [10].

Different from the mentioned literatures using UWB technology, in this paper we design a wireless network for industrial applications based on IEEE 802.15.4a which using chrip spread spectrum technology. We attempt to make a preliminary performance study via serval sets of practical experiments, including the effects of 1) numeral wireless nodes, 2) numeral data packets, 3) data transmissions with different high-layer protocol.

The organization of the paper is as follows. Section 2 introduces the IEEE 802.15.4a communication protocols. Next, experimental hardware and configuration are illustrated in Section 3. Then, experimental results of the performance study are described in Section 4. Finally, Section 5 gives the conclusion and future work.

2 IEEE 802.15.4a Wireless Protocol

This section provides a brief introduction of the standard, focusing on the aspects more relevant to the present work. A more detailed description of the standard can be found in [8, 9, 11]. Medium access within a PAN is controlled by the PAN coordinator that may choose either beacon-enabled or non-beacon-enabled modality.

In this case, two data transfer modes are available:

1) Transfer from a device to the coordinator- a device willing to transfer data to the coordinator uses either ALOHA or slotted Carrier Sensing Multiple Access with Collision Avoidance (CSMA-CA) to access the medium.
2) Transfer from the coordinator to a device- when the coordinator has data pending for a device, it announces so in the beacon. The interested device selects a free slot and sends a data request to the coordinator, indicating that it is ready to receive the data. When the coordinator receives the data request message, it selects a free slot and sends data using either ALOHA or CSMA-CA.

3 Performance Experiments

In this section, we carry out experiments on the network performance of the IEEE 802.15.4a wireless networks.

As shown in Fig. 1, seven nanoNET TRX IEEE 802.15.4a development boards [11] are used as a coordinator and six network devices, respectively, continuously transferring the data so as to perform the measurements.

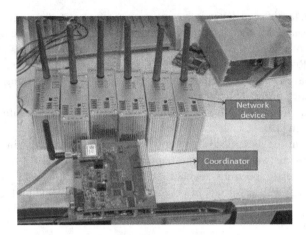

Fig. 1. Experimental equipment

Four sets of experiments are designed to evaluate the various performance of IEEE 802.15.4a, including the effects of 1) numeral wireless nodes, 2) numeral data packets, 3) data transmissions with different high-layer protocol, 4) physical distance between each node. Device 1 is the coordinator continuously sending or receiving data packets to or from the network node. The other Device 2-7 are the network device to send data in the following experiment A-D. The performance study was for a steady state network, i.e. after all the devices finish channel scanning and the relevant procedure to join the PAN.

4   Experimental Results and Discussion

In this section, the time-delay is carried out under the mentioned experimental sets. In these experiments, the time-delay is defined as the time slot between coordinator received signals from the same network device:

$$t_{time-delay} = t_{n,k+1} - t_{n,k} \qquad (1)$$

where n means the nth network device, k means the kth time the coordinator accessing the nth device.

4.1   Effects of Numeral Wireless Nodes

In this experiment, we change the numbers of wireless nodes to observe the effect on time-delay with the token-ring protocol which was designed by our group [12]. First of all, we test the time-delay with only one network device, then add the network device one by one, until the total number of network device is six. Each test has been lasting for 15 minutes at least. All the wireless nodes send only one packet when they get the token-ring to access the wireless channel. For some industrial applications, the communication traffic is very low, for example, the value of temperature or pressure. The data size of packet is set to 168 bytes [11]. The results are shown in Fig. 2.

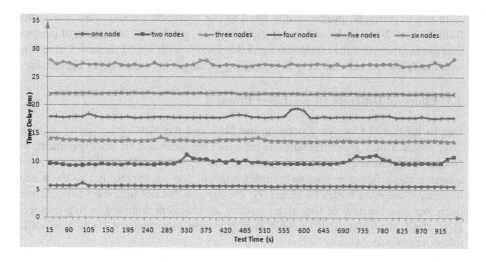

Fig. 2. Effects of numeral wireless nodes

We can see in Fig. 2 that the time-delay is increasing with the number of wireless nodes. The value of increasing changes a few when adding one node, expect the first one. In order to research the law of time-delay variety, we calculate the average value of the time-delay from the experiment result.

4.2 Effects of Numeral Data Packets

During the CSMA/CA, data packets may be undeliverable as the channel is extremely busy, or be erroneous, or even be lost. In this token-ring protocol, we adopt the mechanism of retransmission when the data packet lost until success. So the rate of packet lost is zero. In this subsection, the effect of numeral data packets is studied.

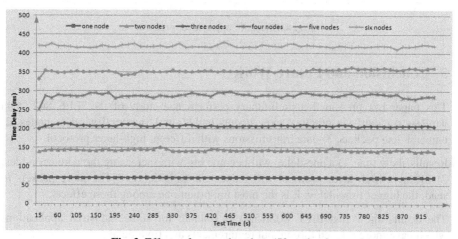

Fig. 3. Effects of numeral packets (50 packets)

In this experiment, packet size is fixed to 168 bytes. 50 packets were transmitted from Devices 2-7 to the coordinator, respectively. Fig.3 showed the effects of numeral data packets between the coordinator and the devices with varied packets. Comparing with Fig.2, we can see that with the increasing of the packets, time-delay became big.

4.3 Effects of Different Upper-Layer Protocol

Due to IEEE 802.15.4a only defining the PHY and the MAC layer, different high-layer protocol may affect the wireless network's performance. In this experiment, we compared different protocol with numeral wireless nodes. Fig.4 shows the result adopting master/slave architecture. In this protocol, the coordinator is also the master to manage the slave nodes (devices2-7). We can see that not only the value of time-delay but also the law of time-delay variety are different from token-ring protocol. First of all, the value of time-delay is bigger than experiment 4.1 which adopting token-ring protocol.

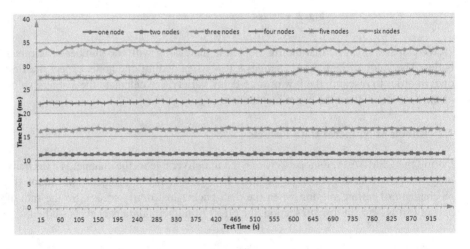

Fig. 4. Effects of upper-layer protocol

5   Conclusion and Future Work

This paper has presented a preliminary performance study of the IEEE 802.15.4a wireless standard via practical experiments including 1) numeral wireless nodes, 2) numeral data packets, 3) data transmissions with different high-layer protocol.

Results show that: 1) Time-delay will become big with increasing the wireless nodes; 2) Efficient high layer protocol will reduce the time-delay; 3)The more packets to send, the bigger time-delay. It is the most important one among these effects;

The results show that IEEE 802.15.4a is suitable for some industrial applications which have more relaxed throughput requirements and time-delay.

Acknowledgements

This work is supported by National Nature Science Foundation of China under Grant 60774059, 60834002 and National High Technology Research and Development Program of China under Grant 2007AA04Z174, 2006AA04030405.

References

[1] Cena, G., et al.: Evaluation of response times in industrial WLANs. IEEE Transactions on Industrial Informatics 3(3), 191–201 (2007)

[2] Willig, A., Matheus, K., Wolisz, A.: Wireless technology in industrial networks. Proceedings of the IEEE 93(6), 1130–1151 (2005)

[3] Tajika, Y., et al.: Networked home appliance system using Bluetooth technology integrating appliances control/monitoring with Internet service. IEEE Transactions on Consumer Electronics 49(4), 1043–1048 (2003)

[4] IEEE 802.15.4a-2007, Part 15.4: Wireless medium access control (MAC) and physical layer (PHY) specifications for low-rate wireless personal area networks (WPANs); Amendment 1: Add alternate PHYs (March 2007)

[5] Choi, J.D., Stark, W.E.: Performance of ultra-wideband communications with suboptimal receivers in multipath channels. IEEE Journal on Selected Areas in Communications 20(9), 1754–1766 (2002)

[6] Weisenhorn, M., Hirt, W.: Robust noncoherent receiver exploiting UWB channel properies. In: Proc. Joint UWBST and IWUWBS, Kyoto, Japan, vol. 2, pp. 156–160 (May 2004)

[7] Tian, Z., Sadler, B.M.: Weighted energy detection of ultra-wideband signals. In: 2005 IEEE 6th Workshop on Signal Processing Advances in Wireless Communications (2005)

[8] D'Amico, A.A., Mengali, U., Arias-de-Reyna, E.: Energy-Detection UWB Receivers with Multiple Energy Measurements. IEEE Transactions on Wireless Communications 6(7), 2652–2659 (2007)

[9] De Nardis, L., Di Benedetto, M.G.: Overview of the IEEE 802.15.4/4a standards for low data rate Wireless Personal Data Networks. In: 4th Workshop on in Positioning, Navigation and Communication, WPNC 2007 (2007)

[10] IEEE 802.15.TG4a official web page,
http://www.ieee802.org/15/pub/TG4a.html

[11] Nanotron: nanoNET Chirp Based Wireless Networks—White Paper. 2005 Nanotron Technologies GmbH (2005)

[12] Hou, W.Y., Yang, A.L.: Protocol of Wireless Monitoring and Control Network based on IEEE802.15.4a. Computer Engineering 35(16), 101–106 (2009)

A Hybrid Ant Colony Optimization and Its Application to Vehicle Routing Problem with Time Windows

Xiangpei Hu[1], Qiulei Ding[1], and Yunzeng Wang[2]

[1] Institute of Systems Engineering, Dalian University of Technology,
Dalian, China, 116023
drhxp@dlut.edu.cn
[2] A. Gary Anderson Graduate School of Management, University of California,
Riverside, California, USA, 92521
Yunzeng.wang@ucr.edu

Abstract. The Ant Colony Optimization (ACO) is a recent meta-heuristic algorithm for solving hard combinatorial optimization problems. The algorithm, however, has the weaknesses of premature convergence and low search speed, which greatly hinder its application. In order to improve the performance of the algorithm, a hybrid ant colony optimization (HACO) is presented by adjusting pheromone approach, introducing a disaster operator, and combining the ACO with the saving algorithm and λ-interchange mechanism. Then, the HACO is applied to solve the vehicle routing problem with time windows. By comparing the computational results with the previous literature, it is concluded that the HACO is an effective way to solve combinatorial optimization problems.

Keywords: Ant Colony Optimization; Vehicle Routing Problem with Time Windows; Combinatorial Optimization Problems.

1 Introduction

ACO, inspired from the foraging behavior of ant species, is developed by A. Colorni et al. [1] in the 1990s. Compared with the previous heuristics, the ACO possesses the characteristics of positive feedback, distributed computing, and easily combining with other heuristics. Recently, the ACO has been shown to be an efficient algorithm for solving combinatorial optimization problems [2, 3].

However, there are some weaknesses of the ACO in dealing with combinatorial optimization problems. Firstly, the search always gets trapped in local optimum. Secondly, it needs a lot of computational time to reach the solution. In order to avoid these weaknesses, T. Stützle et al. [4] presented a MAX-MIN ant system. By adjusting pheromone, the algorithm prevented the search from becoming trapped in local optimum. However, the search speed was influenced since the adjusting pheromone required lots of time. B. Bullnheimer et al. [5] introduced an improved ACO to solve vehicle routing problems. This succeeded at improving the search speed but there was only a slight improvement in solutions. M. Reimann et al. [6] put forward a Divide-Ants algorithm, which was to divide the problem into several disjointed

K. Li et al. (Eds.): LSMS / ICSEE 2010, Part I, CCIS 97, pp. 70–76, 2010.

sub-problems. This algorithm had great advantages when it was used to solve large-scale problems, but its search process was too complicated to prevent its extended application. J. E. Bell et al. [7] proposed an improved ACO combined with the 2-opt heuristic and a candidate list. The search speed of this algorithm was faster, but when it was used to solve large-scale problems, the solutions were worse. A. Aghaie et al. [8] described an improved ACO, in which the obtained solutions were improved but the efficiency solving large-scale problems wasn't explicit.

It is clear that great efforts have been made to improve the ACO. But the premature convergence and low search speed are still ready to be solved. Therefore, this paper tries to provide an improved ACO, namely, HACO, to overcome those mentioned difficulties. The remainder of this paper is organized as follows. Firstly, section 2 describes the principles of the HACO. Secondly, section 3 constructs the model of VRPTW, describes the steps for solving VRPTW and then analyzes the computational results. Finally, section 4 provides conclusions and directions for future research.

2 The Improvement of ACO

2.1 The Approach of Adjusting Pheromone

In consideration of the importance of pheromones between ant colonies, this section focuses on three aspects of adjusting pheromone as follows:

(1) In the ACO algorithm, the pheromone given by ant colonies does not always indicate the optimal direction, and the pheromone deviated from optimal solution has the potential to be enhanced, which prevents the rest of ants from finding a better solution. It is realized that due to the influence of positive feedback, the random choice of the parameters used in the ACO is not good enough to prevent the search from getting trapped in local optimum. Therefore, definite and random selection must be combined with the ACO to improve the global optimization capability, which is carried out by adjusting the pheromone and enhancing the random selection probabilities under the circumstances of the determined evolutionary direction.

(2) At every edge, the maximum or minimum pheromone trails may lead to premature convergence of the search during the process of pheromone updating. Therefore, the HACO puts forward the τ_{min} and τ_{max} as the minimum and maximum pheromone trails. Meanwhile, the pheromone trails are deliberately initialized to τ_{max}, which helps to achieve higher level exploration of solutions at the beginning of the search. Additionally, in cases where the pheromone trails differ greatly, the idea of computing average pheromone trails between τ_{ij} and τ_{max} is absorbed, which will play a significant role in obtaining a new solution [9].

(3) It is difficult for the ACO algorithm to solve large-scale problems because of the existence of the trail evaporation $1-\rho$. If $1-\rho$ is convergent to zero, the global optimization capability will decline because the edges may be chosen repeatedly. The larger $1-\rho$ is, the better the global optimization capability will be. But if so, the convergence speed of the algorithm will be slowed down. Therefore, this paper suggests that a dynamic $1-\rho$ value rather than a constant value is adopted.

2.2 The Introduction of the Disaster Operator

In order to prevent the ACO from getting trapped in local optimum, the disaster operator changing the local optimal solution randomly is introduced. The design of the disaster operator is similar to the mutation of the genetic algorithm. By greatly decreasing pheromone trails in some parts of local optimization routes, the algorithm is able to avoid premature convergence. The routes of disasters are decided by small random probabilities, since the distribution of the pheromone in the previous route would be destroyed by too many occurrences of disasters, which increase the probability of leading the search results to the opposite direction.

2.3 Combining ACO with the Saving Algorithm and λ-Interchange Mechanism

ACO is a strong coupling algorithm for the characteristic of combination with other heuristics. So the speed of convergence will be greatly improved by combining with the saving algorithm [10] and λ-interchange mechanism [11].

Moreover, in the ACO algorithm, it will take a long time to compute the transition probabilities of all unsearched nodes. So choosing the nearest node was adopted to enormously improve the convergence speed by computing the transition probabilities of only those nodes nearby the chosen node.

3 Application of HACO to VRPTW

In order to validate the effectiveness of the HACO in dealing with combinatorial optimization problems, it is applied to VRPTW in this section.

3.1 Construction of VRPTW Model

In this paper, the VRPTW is confined to the following conditions. There is one depot, which owns a few homogenous vehicles. The total demand of the customers in one route must be less than or equal to the loading capacity of the vehicle. Each customer is visited only once and all routes begin and end at the depot. The goal is to find a set of routes of punctual arrival and minimum total cost.

In order to construct the model, we must first define the following notations.

n: the number of customers
v_i: when $i=0$, it denotes the depot. Otherwise, it represents the customer
K: the number of vehicles
C_{ij}: the transportation cost from v_i to v_j
G: the loading capacity of the vehicle
q_i: the demand of v_i
x_{ijk}: binary variable, =1 if vehicle k goes from customer v_i to v_j
y_{ik}: binary variable, =1 if v_i is served by vehicle k
$[ET_i, LT_i]$: the time window of v_i
t_i: the time at which the vehicle arrives at v_i

Then the mathematical model is obtained below:

$$MinZ = \sum_{i=0}^{n}\sum_{j=0}^{n}\sum_{k=1}^{K} C_{ij} x_{ijk} \tag{1}$$

Subject to:

$$\sum_{i=1}^{n} q_i y_{ik} \leq G \quad k=1,2,......,K \tag{2}$$

$$\sum_{k=1}^{K} y_{0k} = K \tag{3}$$

$$\sum_{k=1}^{K} y_{ik} = 1 \quad i=1,2,......,n \tag{4}$$

$$\sum_{i=1}^{n} x_{i0k} = 1 \quad k=1,2,......,K \tag{5}$$

$$\sum_{i=0}^{n} x_{ijk} = y_{jk} \quad j=1,2,......,n; \ k=1,2,......,K \tag{6}$$

$$\sum_{j=0}^{n} x_{ijk} = y_{jk} \quad i=1,2,......,n; \ k=1,2,......,K \tag{7}$$

$$ET_i \leq t_i \leq LT_i \quad i=1, \ 2, \ ......,n \tag{8}$$

In this model, the objective function (1) is to minimize the total cost of routes. Constraint (2) ensures that the total demand of each vehicle route does not exceed the loading capacity of the vehicle. Constraint (3) ensures that all vehicle routes begin at the depot. Constraint (4) ensures that every customer is visited exactly once by exactly one vehicle and that all customers are visited. Constraint (5) ensures that all vehicle routes end at the depot. Constraints (6) and (7) show the relation of variables. Constraint (8) ensures that customers are served within required time.

3.2 Solution Steps of HACO

The steps for solving VRPTW of HACO can be described as follows.

Step 1: Initialize every controlling parameter, assume the optimal solution L_{global}, define the repeated counter as nc, put m ants on the depot, and make a candidate list.

Step 2: Select the next node j to be visited according to formula (9):

$$j = \begin{cases} \arg\max_{j \notin tabu_k} [\tau_{ij}(t)]^{\alpha}[\eta_{ij}(t)]^{\beta}[\mu_{ij}]^{\gamma} & , \quad if \quad q \leq p_t \\ randow \quad j \notin tabu_k & , \quad otherwise \end{cases} \tag{9}$$

Where $tabu_k(k=1,2,...,m)$ is the tabu table which records all the visited nodes by ant k. τ_{ij} and η_{ij} represent the density of pheromones and visibility (the reciprocal of distance d_{ij} between two nodes) respectively. $\mu_{ij}=d_{i0}+d_{0j}-d_{ij}$ is the saving value in the absorbed saving algorithm. α, β and γ are the relative importance of every variable. q is a value chosen randomly with uniform probability in the range [0,1]. p_t ($0<p_t<1$) is initiated as $p_0=1$ and is dynamically adjusted with the evolutionary process.

Step 3: If the total amount of searching ants is smaller than m, go back to step 2 and the remaining ants will continue searching; otherwise, go to step 4.

Step 4: Compute the search solution L_k of every ant, set $L_{local} = \sum_{k=1}^{m} L_k$ as the local optimum solution and save the routes table of L_{local}.

Step 5: 2-interchange mechanism is carried out for z times (z is given a larger initial value in order to completely search the solution space). Update L_{local}.

Step 6: Choose the routes with disasters and set the pheromone trails in these routes equal to minimum pheromone, then compute the solution and update L_{local} again.

Step 7: All the pheromone trails are dynamically updated as follows:

$$\tau_{ij}^{new} = \rho \tau_{ij}^{old} + \Delta \tau_{ij}$$

$$\Delta \tau_{ij} = \begin{cases} Q / L_{local} & , \quad ij \quad belongs \quad to \quad L_{local} \\ 0 & , \quad\quad\quad otherwise \end{cases} \tag{10}$$

Where Q is a constant related to the pheromone trails laid by ants. $1-\rho$ $(0<1-\rho<1)$ interpreted as trail evaporation is initiated as $1-\rho=0$ and is dynamically adjusted with the evolutionary process. After pheromone trails are dynamically updated, τ_{ij} is replaced by τ_{max} when $\tau_{ij} > \tau_{max}$, or by $(\tau_{min} + \tau_{max})/2$ when $\tau_{ij} < \tau_{min}$.

Step 8: If $L_{local} < L_{global}$, set $L_{global} = L_{local}$ and update the routes table simultaneously.

Step 9: As for the complete search of the solution space, the p_t and ρ can be adaptively adjusted as formula (11) and (12).

$$p_t = \begin{cases} 0.95 p_{t-1} & , \quad if \quad 0.95 p_{t-1} \geq p_{min} \\ p_{min} & , \quad\quad otherwise \end{cases} \tag{11}$$

Where p_{min} is the minimum defined in evolutionary process. It is used to ensure the achievement of the definite choosing chance even if the p_t is too small.

$$\rho_n = \begin{cases} 0.95 \rho_{n-1} & , \quad if \quad 0.95 \rho_{n-1} \geq \rho_{min} \\ \rho_{min} & , \quad\quad otherwise \end{cases} \tag{12}$$

Where ρ_{min} is the minimum defined in the evolutionary process for the prevention of a too slow convergence speed when ρ is too small.

Step 10: If nc is bigger than the maximum, then the flow finishes. Otherwise, go back to step 2, and repeat the above steps.

3.3 Computational Results and Analysis

The HACO is tested by Benchmark problems, which are composed of six different problem types (R1, C1, RC1, R2, C2, RC2) [12]. In this paper, R1-01, C1-07, RC1-01, R2-02, C2-01 and RC2-05 are selected randomly as test data from each type of Benchmark problems. Then, a comparison is made between the optima of ten solutions obtained by the HACO, and the solutions of Genetic Algorithm [13], Local Search [14], Tabu Search Algorithm [14], and ACO in Table 1. Table 2 provides the average computing time needed for each problem.

Table 1. Comparison among different heuristics [13, 14]

Heuristics	Problems	R1-01	C1-07	RC1-01	R2-02	C2-01	RC2-05
GenSAT [13]	Vehiles	18	10	14	4	3	4
	Distance	1644	829	1669	1176	591	1389
LS-DIV [14]	Vehicles	19	10	14	6	3	6
	Distance	1648.86	828.937	1677.68	1147.53	591.557	1463.7
SATabu [14]	Vehicles	19	10	14	6	3	6
	Distance	1655.03	828.937	1677.93	1077.66	591.557	1426.09
ACO	Vehicles	19	10	14	6	3	5
	Distance	1702	860	1789	1242	643	1517
HACO	Vehicles	18	10	14	6	3	4
	Distance	1611	831	1126	1130	593	1350

Table 2. The average computing time needed for each problem

problems	R1-01	C1-07	RC1-01	R2-02	C2-01	RC2-05
Time(s)	65	49	57	59	46	53

The platform for conducting the experiments is a PC with a 2.40 GHz CPU and 256 MB RAM. Comparisons in computing time are not made because the platform in this paper is different from the ones in literatures [13, 14]. However, the conclusion obtained from Table 2 shows that HACO spends approximately 60 seconds solving VRPTW with 100 customers and satisfies the need for real-time in distribution.

Table 1 indicates that the solutions of the HACO are better than the solutions of the ACO. Furthermore, several solutions of the HACO are more accurate than the ones of previous literature, while the remainders are very close. It is noted that the solutions obtained are not the optimal ones because the parameters of α, β, Q etc. are set by experiences when the algorithm runs. If every parameter in the algorithm was set at the optimal value, the final solution would be improved further.

4 Conclusions

In order to overcome the weaknesses of the ACO in solving combinatorial optimization problems, this paper has presented the HACO by adjusting pheromone approach, introducing the disaster operator, and combining the ACO with the saving algorithm and λ-interchange mechanism. At the beginning, the premature convergence is avoided effectively and the feasible solution is obtained more easily by adjusting pheromone approach. The unique design of disaster operator also prevents the HACO from getting trapped in local optimum. Then, by combining with the saving algorithm and λ-interchange mechanism, the search speed is greatly improved. The HACO is validated to be an effective way to solve the combinatorial optimization problems by analyzing the computational results.

Possible future research may focus on detailed investigation of parameter values to improve the solution and enhancing the learning ability of ants.

Acknowledgment

This work is partially supported by the grants from the National Natural Science Funds for Distinguished Young Scholar (No. 70725004) and Natural Science Foundation of China (No. 70890080, 70890083, 70571009, 70671014).

References

1. Colorni, A., Dorigo, M., Maniezzo, V.: Distributed Optimization by Ant Colonies. In: Proceedings of European Conference on Artificial Life, Paris, France, pp. 134–142 (1991)
2. Anghinolfi, D., Paolucci, M.: A New Ant Colony Optimization Approach for the Single Machine. International Journal of Operations Research 5(1), 44–60 (2008)
3. Socha, K., Dorigo, M.: Ant Colony Optimization for Continuous Domains. European Journal of Operational Research 185(3), 1155–1173 (2008)
4. Stützle, T., Hoos, H.H.: Improvements on the Ant System: Introducing the MAX-MIN Ant System. In: Artificial Neural Networks and Genetic Algorithms, pp. 245–249. Springer, Heidelberg (1998)
5. Bullnheimer, B., Hartl, R.F., Strauss, C.: An Improved Ant System Algorithm for the Vehicle Routing Problem. Annals of Operations Research 89, 319–328 (1999)
6. Reimann, M., Doerner, K., Hartl, R.F.: D-Ants: Savings Based Ants Divide and Conquer the Vehicle Routing Problem. Computers & Operations Research 31(4), 563–591 (2004)
7. Bell, J.E., McMullen, P.R.: Ant Colony Optimization Techniques for the Vehicle Routing Problem. Advanced Engineering Informatics 18(1), 41–48 (2004)
8. Aghaie, A., Mokhtari, H.: Ant colony optimization algorithm for stochastic project crashing problem in PERT networks using MC simulation. The International Journal of Advanced Manufacturing Technology 45, 1051–1067 (2009)
9. Stützle, T., Hoos, H.H.: MAX-MIN Ant System. Future Generation Computer Systems 16(8), 889–914 (2000)
10. Clarke, G., Wright, J.W.: Scheduling of Vehicles from A Central Depot to A Number of Delivery Points. Operations Research 12(4), 568–581 (1964)
11. Osman, I.H.: Metastrategy Simulated Annealing and Tabu Search Algorithms for the Vehicle Routing Problem. Annals of Operations Research 41, 421–451 (1993)
12. Solomon Benchmark Problems,
    http://www.idsia.ch/~luca/macs-vrptw/problems/welcome.htm
13. Thangiah, S.R., Osman, I.H., Sun, T.: Hybrid Genetic Algorithm, Simulated Annealing and Tabu Search Methods for Vehicle Routing Problems with Time Windows. Technical Report, Institute of Mathematics & Statistics, University of Kent, Canterbury, UK (1994)
14. Tan, K.C., Lee, L.H., Ou, K.: Artificial Intelligence Heuristics in Solving Vehicle Routing Problems with Time Window Constraints. Engineering Applications of Artificial Intelligence 14, 825–837 (2001)

Modeling and Simulation of a Yacht Propulsion System

Yihuai Hu, Xiaoming Wang, and Huawu Zhang

Department of marine engineering, Shanghai Maritime University, Shanghai 200135
yhhu@shmtu.edu.cn

Abstract. This paper firstly introduces the schematic diagram of a yacht propulsion system. Mathematical models of the yacht propulsion system are then proposed including main diesel engine, reduction gearbox, hydraulic clutch and propeller. The programming of simulation software and the design of simulation hardware are described. The practical operation with this training software is also introduced, which could be used for operation skill training and certificate assessment of Yachtsmen.

Keywords: yacht; propulsion system; modeling and simulation.

1 Introduction

In America, Every ten people own one yacht on average. The total number of Yachts every year has reached to 18.5 million, as well as the annual sales of 300 billion dollars. As a kind of entertainment tool, yachts are quite popular in the developed countries. With the development of national economy and the improvement of people's living standard in China, requirement for yachts has been greatly demanded in the market. In recent years, those investors who once invested public parks and major tourism attractions begin to invest heavily in developing water-on tourism and yacht projects. At the same time, some rich businessmen and some middle class people want to own a private yacht. Some yacht clubs are established in big cities. Yacht industry gets a historic opportunity for the development in China. It is estimated that yacht industry in China is about one hundred billion dollars.

As an entertainment tool the yacht is relative safe. But it can be very dangerous without correct operation on water. According to the regulation by China Maritime Safety Bureau, yachtsmen must get driver's license after training before practical operation on the water. Yacht simulator here could play an important role in improving the driver's ability in practical operation. Therefore, the study on simulation of the yacht is particularly essential.

2 Modeling of Propulsion System

A main propulsion shafting system (including the propeller) consists of the equipment necessary to convert the rotative power output of the main propulsion engines into thrust horsepower, suitable for propelling the ship, and the means to impart this thrust to the

K. Li et al. (Eds.): LSMS / ICSEE 2010, Part I, CCIS 97, pp. 77–82, 2010.

ship's hull. There are several kinds of marine propulsion system as direct drive system, indirect drive system, controllable pitch propeller system and electrical drive system. The yacht propulsion system introduced in this paper is a kind of twin-engine, twin-propeller indirect drive system installed on a "Pu Jiang" yacht. It includes two main engines, reduction gearboxes, transmission shaft and two propellers as shown in Fig. 1.

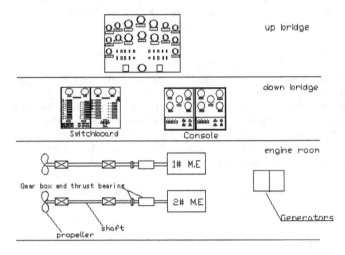

Fig. 1. Schematic diagram of yacht propulsion system

2.1 Main Engine and Propeller

The simulation models for main diesel engine is of filling emptying type, which treats a multi-cylinder engine as a series of thermodynamic control volumes interconnected through valves and ports. A three-zone scavenging mode with fresh air, exhaust gas and mixed zones was used. Turbocharger compressor and turbine experimental performance maps were included in a digitised form and then interpolated within the data to find the operating point. The main diesel engines used on the yacht are two high-speed diesel engines. A real time simulation algorithm of main engine was made based on the traditional filling emptying diesel engine model. Within the possible running range of main engine, running speeds n_1, n_2, ..., n_{10} and fuel racks S_1, S_2, ...S_{10} were selected and thermodynamic variables under each running speed n_i and each fuel rack S_i ($i = 1,2,...,n$) were calculated with filling emptying method which formed a variable matrix A.

$$A = \begin{array}{c} \\ n_1 \\ n_2 \\ \cdot \\ \cdot \\ \cdot \\ n_m \end{array} \begin{array}{cccccc} Y_1 & Y_2 & & & & Y_n \\ \left[a_{1,1} \right. & a_{1,2} & \cdot & \cdot & \cdot & a_{1,n} \\ a_{2,1} & a_{2,2} & \cdot & \cdot & \cdot & a_{2,n} \\ \cdot & \cdot & & & & \cdot \\ \cdot & \cdot & & & & \cdot \\ \cdot & \cdot & & & & \cdot \\ a_{m,1} & a_{m,2} & \cdot & \cdot & \cdot & \left. a_{m,n} \right] \end{array}$$

(1)

(m=10, n=10)

Assume the present running speed is n and actual fuel rack is S during simulator's training, thermodynamic variables in vector B under running speed n and 10 fuel racks S_i (i=1,2, ...,10) are firstly got with Newton interpolation method as

$$B = \left\{ \begin{matrix} s_1 & s_2 & . & . & . & s_n \\ b_1 & b_2 & . & . & . & b_n \end{matrix} \right\} \tag{2}$$

$$c = \frac{(s-s_{i+1})(s-s_{i+2})}{(s_i-s_{i+1})(s_i-s_{i+2})} b_i + \frac{(s-s_i)(s-s_{i+2})}{(s_{i+1}-s_i)(s_{i+1}-s_{i+2})} b_{i+1} + \frac{(s-s_i)(s-s_{i+1})}{(s_{i+2}-s_i)(s_{i+2}-s_{i+1})} b_{i+2} \tag{3}$$

Where, $S_i < S < S_{i+1}$

By this algorithm thermodynamic variables of the diesel engines under any performance condition could be obtained with only eleven interpolating calculations and without plenty of iteration calculation of differential equation.

Assume the new running speeds of diesel engines are n'1 n'2 respectively and the new navigating speed of vessel is v_s', then

$$n_1' = n_1 + (M_{e1} - M_{p1} \cdot |D_{r1}| - M_{f1} \cdot \text{sgn}(n_1)) / 2 / \pi / Je_1$$
$$n_2' = n_2 + (M_{e2} - M_{p2} \cdot |D_{r2}| - M_{f2} \cdot \text{sgn}(n_2)) / 2 / \pi / Je_2 \tag{4}$$
$$v_s' = v_s + (T_1 \cdot \text{sgn}(D_{r1}) + T_2 \cdot \text{sgn}(D_{r2}) - R \times \text{sgn}(v_s)) / m / (1 + 0.06)$$

Where:

$\qquad M_{e1}, M_{e2}$: Main engine torque moments (N.m);

$\qquad M_{p1}, M_{p2}$: Propeller rotating torque (N.m);

$\qquad M_{f1}, M_{f2}$: Propeller resistant moment (N.m);

$\qquad J_{e1}, J_{e2}$: Rotating inertia of two shafts (kg.m^2);

$\qquad T_1, T_2$: Propeller propulsive forces from two main engines (N);

$\qquad D_{r1}, D_{r2}$: Running directions of two propellers;

$\qquad R$: Ship drag force (N);

$\qquad m$: Ship mass (kg).

According to the vessel trial, the propeller rotating torque Mp, propeller propulsive force T, ship drag force R, propeller consuming power Pp and the engine indicated power P_i could be got as

$$M_p = C_1 K_M n_p^{\ 2} \tag{5}$$

$$T = C_2 K_T n_p^{\ 2} \tag{6}$$

$$R = C_3 v_s^2 \cdot Y_R \tag{7}$$

$$P_p = C_4 K_M n_p^3 \cdot Y_R \qquad (8)$$

$$P_i = P_p /(\eta_m \eta_{rt} \eta_w) = C_5 K_M n_p^3 \cdot Y_R \qquad (9)$$

Here Y_R is a gain factor, which could express different navigation conditions such as hull cleanliness, wind scale, rudder angle, navigation channel and ship draft [1].

2.2 Hydraulic Clutch

The function of the hydraulic clutch is to connect the main engine crankshaft and propeller shaft together and transfers the power from ain engines to the propellers when necessary. Before engagement the clutch could be divided into two parts and each part has different rotating speed n_e and n_p respectively.

$$M_e - M_c - M_{f1} = 2\pi J_1 \tfrac{dn_e}{dt} \qquad (10)$$

$$M_c - M_{f2} - M_p = 2\pi J_2 \tfrac{dn_p}{dt} \qquad (11)$$

Where: Mc——Clutch driving torque;

M_{f1}, M_{f2}——Clutch mechanical resisting torques;

J_1, J_2——Rotating inertia of engine shaft and propeller shaft.

After clutch engagement the rotating speed of propeller can be as:

$$n_p = K_C \cdot i_c \cdot n_e \qquad (12)$$

Where: K_C——Clutch status coefficient;

1——engaged of clutch

0——disengaged of clutch

i_c——Speed reduction ratio

3 Simulation Interface Design

The Yacht propulsion simulation system consists of five software human-machine interfaces and five hardware operation boxes. The software interfaces include bridge control panel, main engine control boxes, 220V AC power switchboard, 24V DC power distribution box and a diesel generator starter. The simulation hardware includes 1# main engine control box, 2# main engine control box, 220V AC power switchboard, 24V DC power distribution box and a diesel generator local starter. These software interfaces were designed on Visual Basic 6.0 platform in Windows'2000 operating system and the interfaces could dynamically display operating parameters of the propulsion system. The bridge operation interface consists of indicators of engine fuel rack, engine speed, lubricant oil pressure, cooling water temperature, propeller speed and boat speed. There are also two simulated clutch operating handles and two main engine fuel rack levers as shown in Fig. 2. The simulated 24V DC power distribution box is shown in Fig. 3 with current meter, voltage meter, power switches and alarm indicators.

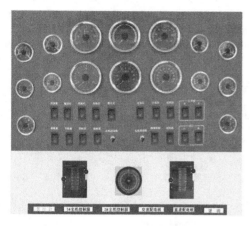

Fig. 2. The bridge control software interface   **Fig. 3.** The DC power distribution box

Any operation on hardware will be converted into a digital signal or analog signal, and then inputted into the simulation computer by data acquisition card. Pushbuttons, switches and other input signals could be entered into the computer system through ISA / PCI bus card. Main engine running, speed, lubricant oil pressure, cooling water temperature, propeller rotating speed and yacht navigating speed are then calculated by the mathematical models mentioned above and outputted to the hardware indicators, which develop the real responses and performances of the simulated yacht propulsion system.

4 Conclusions

By now, this simulation system has been successfully putted into practical training for yachtsmen. It has proved that the simulator training could significantly enhance driver's operating skill and get good economic and social benefits for the users.

References

1. Hu, Y., Shi, W., Huang, X.: Simulation modeling of engine room simulator for river-going vessel. In: Proc. of 5th International Conference on System Simulation and Scientific Computing, Shanghai, pp. 457–462 (2002)
2. Xu, L., Zhan, Y.: Marine power plant technology management. Dalian Maritime University Press, Dalian (2006)

3. Guo, Q.: System Modeling Theory and Method. National Defense University Press, Changsha (2003)
4. Zhuang, Z.: Port of new yacht transport boat and advanced power plant design. In: Ships, pp. 34–38 (1996)
5. Yang, X.: Some thoughts on the development of China's yacht. In: China Ship Survey, pp. 31–32 (2000)
6. Jiang, W., Shang, J., et al.: High-speed yacht Driving Simulator Research and Design. Jiangsu Science and Technology University (Natural Science Edition), 13–15 (2005)
7. Sun, J., Guo, C.: Large-scale marine power plant modeling and simulation. Journal of System Simulation, 465–469 (2007)

Two-Phase Clock Auction Design

Lanbo Miao[1,2] and Jiafu Tang[1]

[1] Dept of Systems Engineering, College of Information Science & Engineering,
Northeastern University, Shenyang, 110819, P.R. China
[2] China United Network communications Corpration Liaoning Branch,
Shenyang, 110002, P.R. China
lbmiao@mail.sy.ln.cn, jftang@mail.neu.edu.cn

Abstract. We propose the two-phase clock auction as a practical means for auctioning many units, a private values clock phase is followed by an interdependent values clock auction phase. The approach combines the simple and transparent price discovery of the private values clock auction with the efficiency of interdependent values clock auction. The private values clock phase is maintained as long as possible to speed up the auction process, and then is taken over to an interdependent values clock auction to improve efficiency and enhance sellers' revenues.

Keywords: Clock auction, Private values, Interdependent values, Efficiency.

1 Introduction

Our two-phase clock auction is a hybrid auction format that begins with private values clock auction phase and ends with an interdependent values clock auction phase. First, bidders directly submit bids in a clock auction with private values. Just before excess supply appears, the clock auction with interdependent values begins. The clock time for the two phases is the same.

The traditional clock auction is an iterative auction with private values in which the auctioneer announces price and total quantity, all bidders indicate the quantities of units desired at the current prices, then the price for units with excess demand increases, the bidders again express quantities at the new prices. This process is repeated until there are no units with excess demand.

Vickrey [1] sets up a static multi-unit auction and the auction achieves an ex-post efficient outcome with pure private values. Ausubel [2] extends a modified second-price auction defined by Maskin [3] to a dynamic generalized Vickrey auction for multiple identical items with private values. Unfortunately, the Vickrey auction is no longer efficient once we leave the private values setting, Partha Dasgupta and Maskin [4] define a static auction with interdependent values designed to attain efficiency, but the mechanism requires bidders to submit to the auctioneer their entire preference profile of the other bidders, while Perry and Reny [5] establish a two-round interdependent values auction and the auction also requires bidders to submit large amounts of redundant information.

K. Li et al. (Eds.): LSMS / ICSEE 2010, Part I, CCIS 97, pp. 83–91, 2010.

The Ausubbel's private values clock auction [2] is a simpler process, bidders are provided the minimal information needed for price discovery—the prices and the total excess demand, the auction is easy for bidders to take part in and fast to make a bid decision, but it allows the price to fall at last. Perry and Reny [6] study a distinct ascending multi-unit auction with interdependent values further and show that it is efficient under downward-sloping demand, but if there are too many bidders in the auction the auction will take a long time to achieve equilibrium. By a different approach in applying Ausubel and Perry's result and working directly with their main theorem, we establish a two-phase clock auction to overcome the shortcomings above.

2 Auction Mechanism Design

The two-phase clock auction begins with private values clock auction and concludes with an interdependent values clock auction phase. The clock time that publicly displays the current price is discrete in private values clock auction phase while the clock time turns to be continuous in interdependent values clock auction phase.

2.1 Private Values Clock Phase

There are K units of identical goods to be distributed among m bidders. A feasible allocation is $x \equiv (x_1, ..., x_m)$ such that $\sum_i x_i = K$ while bidder $i = 1, ..., m$.

At first, the auctioneer announces the clock price for time $t = 0, 1$ and the number K for total available supply. Every bidder i reports his demand q_i^t for clock time $t = 0, 1$.

From time $t = 1$ on, the auctioneer only announces the clock price for the next time and available supply for current clock price. Every bidder i only reports his demands q_i^{t+1} for the next time. This process repeats until there is no excess supply for the next time according bidders' report.

At any clock time t, after the auctioneer announces the clock price, each bidder bids his demands for the next time. Demands are constrained by some rules below:

Rule 1 (Monotonic rule). $q_i^{t+1} \leq q_i^t$, for all bidders $i = 1, ..., m$ and $t = 0, \cdots\cdots, L$, while L is the clock time for the last round in the first clock phase that satisfies $\sum_i q_i^{t+1} \leq K$.

The rule1 means that bidder i can maintain or decrease the quantity as the clock price rises, but the quantity can not increase.

After announces each clock price, the auctioneer will calculate the cumulative allocation for every bidder till current time by

$$x_i^t = \max\{0, K - \sum_{j \neq i} q_j^t\}. \tag{1}$$

For $t = 0, ..., L$ and $i = 0, ..., m$.

Rule2 (Drop off rule). $x_i(t) \le q_i(t)$, for all bidders $i = 1,...,m$ and $t = 0, \cdots \cdots, L$.

Once $q_i^t < x_i^t$, bidder i will drop off with the cumulative allocation x_i^t and the active bidders left decrease one. This rule means that bidder i can win no more than he bids.

At every time t, the auctioneer will form the aggregate supply $\sum_i q_i^{t+1}$ from the individual bids for time $t + 1$. If there is excess supply by prediction at time $t + 1$, then a new round begins for time $t + 1$. Otherwise, private values clock phase ends and interdependent values clock phase begins.

$$\Delta x_i(t) = x_i^t - x_i^{t-1}. \tag{2}$$

For all time $t = 0,...,L-1$ and each bidder $i = 0,...,m$.

Then the total payment for the private values clock auction phase is

$$y_i^t = \sum_{t=0}^{L} p^t(\Delta x_i^t). \tag{3}$$

While p^t is the pricing for time t.

2.2 End of the Private Values Clock Auction Phase

As soon as $\sum_i q_i(t+1) \le K$, the private values clock phase ends and the round for the current time is assigned to be the last round while the clock time is $L = t$. The auctioneer reports the total supply left $\overline{K} = K - \sum_i x_i^L$ and the cumulative allocation x_i^L of every bidder i for the moment. Because $\sum_i x_i^L$ units have been allocated up on the current time, the demands of bidder i for the interdependent values clock auction phase is initialized to $q_i(t) = q_i^L - x_i^L$ units. All the bidders whose demands are greater or equal to their bids have dropped off, let the active bidders left be $\overline{m}$ units and re-order them by $1,...,\overline{m}$.

Then we set up $\overline{m}$ auctions, each active bidder is responsible for one of them. The quantity for sale in each auction is initialized to $\overline{K}$ and each active bidder bids $\overline{K}$ units in all the others' auctions. The auction clock is changed to be continuous and turns on with the time that the private values price-clock ends, then interdependent values clock phase begins.

2.3 Interdependent Values Clock Auction Phase

During interdependent values clock auction phase, all auctions share the same price clock, the clock will be turned on and off at various points during all the phase at the same time. The auctioneer is responsible for turning the clock on, while the bidders turning the clock off when they reduce their demands. When the price clock is on, the clock price rises continuously at a constant rate. When the price clock is off, it displays the price it reached the moment it was turned off. Consequently, the clock price never decreases. All actions taken by all bidders are public information, so the values information for each bidder is interdependent and almost common.

After the first phase, there are $\overline{K}$ units of goods left to be distributed among $\overline{m}$ bidders. Each bidder privately receives a one-dimensional signal $\theta_i \in \Theta_i \subseteq \mathbb{R}$. The entire vector of signals determines the bidders' marginal values. Let the vector of signals is $\theta = (\theta_1, ..., \theta_{\overline{m}})$, then the marginal value of bidder i for his $k-th$ unit is denoted by $v_{ik}(\theta_i, \theta_{-i}) = v_{ik}(\theta)$ where $\theta_{-i} \in \mathbb{R}^{\overline{m}-1}$ is the vector of signals of all the other bidders. We shall maintain the following assumptions throughout the following paper for all $i, j = 1, 2, ..., \overline{m}$, all $k, l = 1, 2, ..., \overline{K}$ and all $\theta \in \Theta = \theta_1 \times ... \times \theta_{\overline{m}}$.

Assumption 1. $v_{ik}(\theta) \geq v_{ik+1}(\theta) \geq 0$. This assumption says that marginal values for the bidders are weakly decreasing in k.

Assumption 2. $\partial v_{ik}(\theta) / \partial(\theta_i) > 0$. This assumption is a monotonic condition, says that $v_{ik}(\theta)$ is strictly increasing in θ_i.

Assumption 3. $\partial v_{ik}(\theta) / \partial(\theta_i) > \partial v_{jl}(\theta) / \partial(\theta_i)$, while $v_{ik}(\theta) \geq v_{jl}(\theta)$ for all $i \neq j$.

This assumption says that the inequality is strict when θ_i rises or θ_j falls and all other components of θ remain unchanged. The assumption is satisfied whenever one's own signal affects his marginal values more than it affects others' marginal values, which in turn the assumption ensures that the bidder with the highest signal also has the highest value. This is a single-crossing property.

Assumption 4. $P_i(\{v_{ik} = x\}) = 0$, for all $i = 1, ..., \overline{m}$ and $k = 1, ..., \overline{K}$. This assumption rules out the particular values occur with positive probability.

Assumption 5. (Full Support Prior) The valuations have positive density over their respective intervals. This guarantees that some bidders will be forced to make bids that are attractive enough to induce other bidders to be active.

Assumptions3, assumption4 and assumption5 together ensure that the marginal values of distinct bidders are distinct with prior probability one.

If any bidder wants to decrease his demands, he just decreases his bids which he bids in the others' auctions at time t and turns the clock off. If more than one bidder reduces his demands at the same time, one of them is chosen at random and only his bids reductions take effect at that time.

A bid-increase process begins just after a bid-decrease process at a particular price. During the bid-increase process, each bidder, one after the other, is given the opportunity to increase his bids in any of the others' auctions while the price clock stops all the time. The two process iterate continuously till no one wants to reduce his demands, that also means no bidder want to decrease his bids, then the auctioneer once again turn the price clock on allowing the price to rise from where it stopped.

Bidder i gets every unit that he demands from his own auction and wins the price for the unit from $j \neq i$'s auction. If bidder i wins a unit, two conditions must be satisfied:

Condition 1. The quantity of bidders- $j \neq i$ who is bidding in $i's$ auction must be low enough so that

$$\bar{K} - x_i(t) - \sum_{j \neq i} q_j^i(t) > 0.$$

While- $x_i(t)$ is the units quantity allocated to bidder i during the interdependent values clock auction phase and $q_j^i(t)$ is the units quantity that bidder $j \neq i$ bids in auction i till the clock stopped just now.

This condition also means that the price should be high enough, so that there are enough units for bidder i to win, because such a high price make enough bidders drop off in i's auction.

Condition 2. The price for each unit that bidder i wins must be bided in the others' auctions instead of his own.

Condition 3. $\sum_{i \neq j} q_i^j(t) > x_i(t)$. The units demands that bidder i is bidding in others' auctions must exceed the quantity $x_i(t)$ that i has already won during this phase.

The auction ends when all units have been won or all bids in all auctions are zero at the end of a bid-increase process.

3 Equilibrium and Implementation Issues

3.1 Strategies for Private Values Clock Auction Phase

At time t, each bidder i may be allocated up to q_i units that bidder i bids, let $U_i(q_i)$ be the total value of bidder i for q_i units and it is assumed to be quasi-linear. In the private values clock auction phase bidder i only knows and cares about himself, so in every clock price bidder i can use his private values function $U_i(.)$ to predict his bid for the next clock price.

Definition 1. Given a clock time t, we define truthful prediction report strategy by

$$\hat{q}_i^{t+1} \equiv \inf\{\arg\max_{q_i \in X_i}\{U_i(q_i) - p^{t+1}q_i\}\}. \tag{4}$$

The truthful prediction report is a strategy for that ensures any bidders to gain the maximum utility with the minimum bids.

3.2 Strategies for Interdependent Values Clock Auction Phase

For all $i, j = 1, 2, ..., \overline{m}$, all $k, l = 1, 2, ..., \overline{K}$ and every vector of signals $\theta \in \Theta$, let

$$
\hat{\theta}_{ik}(\theta_{-j}) = \begin{cases} \inf\{\theta_i \mid v_{ik}(\theta_i, \theta_{-j}) < v_{jl}(\theta_i, \theta_{-j})\} \\ \theta_i \text{ if } v_{ik}(\theta_i, \theta_{-j}) = v_{jl}(\theta_i, \theta_{-j}) \\ \sup\{\theta_i \mid v_{ik}(\theta_i, \theta_{-j}) > v_{jl}(\theta_i, \theta_{-j})\} \end{cases}
\tag{5}
$$

Accordingly, define

$$
v_{ik}^{jl}(\theta_{-j}) = \begin{cases} v_{ik}(\hat{\theta}_{ik}(\theta_{-i}), \theta_{-j}) \text{ if } v_{ik}(\theta_i, \theta_{-j}) \neq v_{jl}(\theta_i, \theta_{-j}) \\ v_{ik}(\theta_i, \theta_{-j}) \quad\quad \text{ if } v_{ik}(\theta_i, \theta_{-j}) = v_{jl}(\theta_i, \theta_{-j}) \end{cases}
\tag{6}
$$

Let B be the subset of all active bidders whose signals are revealed and it contains bidder i and θ_B be the vector of their signals, then we define $\overline{v}_{ik}^{jl}(\theta_B; B) = \min v_{ik}^{jl}(\theta_{B-j})$.

Definition 2. The strategy for the interdependent values clock auction phase is defined: The action that bidder i takes at time t given all history actions is to keep his current bids in every auction $j \neq i$ unless one of the following conditions holds.

(a)During a bid-decrease process and when it is i's turn to bid, if $q_i^j > \overline{q}_i^j$ and current clock price is p, bidder i decreases his bids in auction $j \neq i$ to $\overline{q}_i^j$ units, while $q_i^j = k$ is the bids of bidder i in auction $j \neq i$ and $\overline{q}_i^j = \max\{k \mid \overline{v}_{ik}^{jl}(\theta_B; B) > p\}$.

(b)During a bid-increase process and when it is i's turn to bid, suppose that according to the history till now bidder j's bids in auction i fell from l to l' units at the current clock price p; and the total bids in auction i is $\overline{K} - k$ or less while $k = x_i + 1$. If $\overline{v}_{ik}^{jl}(s_B; B) \geq p$ and bidder i's bids in auction $j \neq i$ is below k for the next unit he bids, then bidder i increases his bids in auction j to k units.

If (a) does not hold and the bids of bidder i in auction $j \neq i$ is less than k, bidder i increases his bids in auction $j \neq i$ to $\overline{q}_i^j$ units.

3.3 The Existence of Equilibrium

Definition 3. The payment rule for the interdependent values clock auction phase is defined that bidder i pays the $(\overline{K}-k+1)-th$ highest values he wins in auction $j \neq i$ for her $k-th$ unit. For a revealed signal vector θ_{-i} of others let $\pi_{ik}(\theta_{-i}) = v_{-i}(\hat{\theta}_{jl}(\theta_{-i}, k), \theta_{-i}, \overline{K}-k+1)$ denote the $(\overline{K}-k+1)-th$ highest member of all $\{v_{jl}^{ik}(\theta_{-i})\}_{j,l}$ where j runs over all auctions and l runs over all units $1,...,\overline{K}$.

We then have the following proposition.

Proposition 1. Consider the interdependent values clock auction phase, the above strategies for the interdependent values clock auction phase are efficient and the phase under these strategies constitutes an ex-post equilibrium under assumptions 1-4.

Proof: See the appendix.

Theorem 1. The two-phase clock auction has an ex-post equilibrium with truth-telling strategies.

Proof: See the appendix.

3.4 Implementation Issues

The redundant information is limited in the auction phases. In the first phase, bidders are provided the minimal information needed for price discovery—the prices and the total excess demand. In the second phase, bidders monitor the past actions of the others to infer the others' private information and straightforwardly bid price for each unit against the auction that may win with the most possibility at current clock price. While simultaneous static auction formats often require bidders to provide significant amounts of redundant information, as if it were costless to collect and costless to provide.

Allocation and pricing is independent from any bidders for all the two phases, truth telling is given priority in the auction mechanism. In the first phase, each bidder's allocation is determined by others' demand while the unit price is determined by the current clock. In the second phase, each bidder gets every unit that he demands from his own auction but his allocation is also determined by others' demand, and the bidder wins the price for the unit from others' auctions.

The clock time is the same one for the two phases, there is no price fall during all the auction phases. By making use of the private value setting in the first phase, we let bidders predict the demand of next clock period. By calculating he difference between demand and supply, we combine the private values and interdependent values clock auctions together, so the clock price never decreases including the last round.

4 Conclusions

We have provided a two-phase clock auction for auctioning many units—a private values clock auction followed by an interdependent values clock auction. We use the private values clock phase as long as possible to speed up the course of price discovery, and by adding interdependent values clock auction phase, the price competition among all the bidders gets improved and the allocation becomes more efficient. Truth telling is given priority in the auction mechanism; the equilibrium for the auction exists and is ex-post by truth telling.

Acknowledgments. This research is supported by National Natural Science Foundation of China (NSFC 70721001 and 70625001).

References

1. Vickrey, W.: Counterspeculation, Auctions, and Competitive Sealed Tenders. J. Journal of Finance 93, 675–689 (1961)
2. Ausubel, L.M.: An Efficient Ascending-Bid Auction for Multiple Objects. J. American Economic Review 94(5), 1452–1475 (2004)
3. Maskin, E.: Auctions and Privatization. In: Siebert, H. (ed.) Honor of Herbert Giersch, Tubingen, JCB Mohr, pp. 115–136 (1992)
4. Dasgupta, P., Maskin, E.: Efficient Auctions. J. Quarterly Journal of Economics 115, 341–388 (2000)
5. Perry, M., Reny, P.J.: An Efficient Auction. J. Econometrica 70(3), 1199–1212 (2002)
6. Perry, M., Reny, P.J.: An Efficient Multi-Unit Ascending Auction. J. Review of Economic Studies 72, 567–592 (2005)
7. Yoon, K.: An Efficient Double Auction. J. Journal of the Korean Econometric Society 17(2), 1–20 (2006)

Appendix

The proof of proposition1 requires the lemma1 from Perry and Reny [5]. Its proof is analogous to the proof (4.2) in Section 4.

Lemma 1. The function $v_{ik}^{jl}(.)$ defined in (6) satisfies the following condition:

If $v_{ik}^{jl}(\theta_{-j}) \geq v_{jl}^{ik}(\theta_{-i})$, then $v_{ik}^{jl}(\theta_{-j}) \geq v_{ik}^{jl}(\theta) \geq v_{jl}^{ik}(\theta) \geq v_{jl}^{ik}(\theta_{-i})$.

The proof of proposition 1: Suppose that bidder i is bidding the price for his $k-th$ unit in auction j. By the fact that $v_{jK}^{ik}(\theta_{-i}) \leq v_{jK-1}^{ik}(\theta_{-i}) \leq ... \leq v_{j1}^{ik}(\theta_{-i})$ and the total bids against i must be no more than $\bar{K}-k$, according to the definition3 for payment, we may conclude that bidder i cannot win his $k-th$ unit for a price less than $\pi_{ik}(\theta_{-i})$, and we can also know that bidder i cannot win a $\bar{K}-th$ unit for a price less than $\pi_{i\bar{K}}(\theta_{-i})$, after finite times of bids reduction and increasing $\pi_{i\bar{K}}(\theta_{-i})$ must reach, this means that all the $\bar{K}$ units can be sold at last and that this lower bound is independent of i's behavior, so a feasible allocation achieves at last.

Suppose the feasible allocation for bidder i is k^*. From the definition of payment rule from definition 3, we can know that bidder i's payoff is

$$\sum_{k=1}^{k_i^*(\theta_i',\theta_{-i})}[v_{ik}(\theta_i,\theta_{-i})-\pi_{ik}(\theta_{-i})]$$
$$=\sum_{k=1}^{k_i^*(\theta_i',\theta_{-i})}[v_{ik}(\theta_i,\theta_{-i})-v_{-i}(\hat{\theta}_{jl}(\theta_{-i}),\theta_{-i},\bar{K}-k+1)]$$
(7)

By the definition $\hat{\theta}_{jl}(\theta_{-i})$ from (5) and values monotonicity, we have

$$v_{ik}(\theta_i,\theta_{-i})-v_{-i}(\hat{\theta}_{jl}(\theta_{-i}),\theta_{-i},\bar{K}-k+1)>0 \text{ for all } k<k^*$$
$$v_{ik}(\theta_i,\theta_{-i})-v_{-i}(\hat{\theta}_{jl}(\theta_{-i}),\theta_{-i},\bar{K}-k+1)=0 \text{ for all } k=k^*$$
(8)
$$v_{ik}(\theta_i,\theta_{-i})-v_{-i}(\hat{\theta}_{jl}(\theta_{-i}),\theta_{-i},\bar{K}-k+1)<0 \text{ for all } k>k^*$$

Therefore, bidder i's payoff is maximized while $\theta_i'=\theta_i$, it is ex-post perfect equilibrium in this phase and truth telling bidding is a weakly dominant strategy in this phase.

The proof of theorem 1: We begin with $t=L$, if $q_i(t+1)>\hat{q}_i(t+1)$, this may cause $K<\sum q_i(t+1)$, and result that the first phase auction won't end at the time $t+1$.

We know from proposition1 that the second auction phase is efficient, so bidder i will unprofitably win units at the time $t+1$ that she could have avoided winning by predicting (bidding) sincerely for time $t+1$, the choice will have no effect on the subsequent bidding by opponents otherwise.

While $t<L$, the standing cumulative quantity of allocated units for bidder i at time t is given by

$$x_i(t)=\max\{0,K-\sum_{j\neq i}q_j(t)\}.$$
(9)

Since the equation is independent of the actions of any bidder, changing one's own bid strategy can have no effect on his payoff. By definition for truthful prediction report strategy, every bidder can gain the maximum utility with truth telling.

By the proposition1 we can know that the interdependent auction phase is efficient, and truth telling bidding is a mutual best response for every bidder. Hence it is an ex-post perfect equilibrium for the two-phase clock auction, and truth telling bidding is a weakly dominant strategy.

Research on Simulation of Multi-agents Competition Model with Negotiation

Liqiao Wu, Chunyan Yu, and Hongshu Wang

College of Mathematic and Computer Science, Fuzhou University
350108 Fuzhou, Fujian, China
177511wlq@163.com

Abstract. To construct an artificial system with multi-Agents, it is obvious that some important factors, such as different agent's interests, limited resources and so on, will inevitably lead to conflict. To reduce conflict, it is found that effective competition with negotiation among multi-Agents can improve overall performance. Thence, this paper proposes a new Multi-Agents Competition Model with Negotiation, which improves the forecast accuracy of opponent's competing strategies with negotiation information and shortens negotiation time effectively depending on selecting strategies in probability to maximum interest.

Keywords: Competition with negotiation; Multi-Agents competition model; Negotiation information; Probability selection strategies.

1 Introduction

Multi-agent system (MAS), adopted by most artificial systems, mainly focuses on establishing a self-government system which has the ability to solve problems itself. With the increasing complexity of the actual systems and interaction of subsystem, it has become an urgent task to understand how the agent interact with others, how to organize such large number of agents and how to realize cooperation and negotiation among agents.

Many factors, such as constantly changing environment, limited resources and different interest of agents, will inevitably lead to conflict which will reduce the system performance. In order to avoid these conflicts and achieve the best performance, agents need to compete by means of negotiating. Meanwhile, the initial belief of agents is obviously insufficient to meet the ever-changing environment and agents' selection mechanism of behavior and strategy will directly affect the process of negotiation, so it is crucial to study how to effectively update the agent's belief and perfect the learning mechanism [3].

2 Multi-agents Competition with Negotiation

In MAS, negotiation is a very effective approach to help agents to achieve win-win when they are in competition. The competition with negotiation should include three key elements [1,2]:(1)Competition goal, which is the target that agents want to reach

K. Li et al. (Eds.): LSMS / ICSEE 2010, Part I, CCIS 97, pp. 92–99, 2010.

and should satisfy agents' interest. (2)Negotiation protocol: Negotiation protocol is a set of rules that agents should comply with.(3)Negotiation decision-making model: The process of negotiation also is a process of strategy selecting.

2.1 Competition Goal

Competition goal determines the issues should be involved with when agents carry out negotiation. It has two types: single target and multiple targets. Single target implies one agent's gains increase will result in the others' gains decrease, it is difficult to achieve win-win. While multiple targets denote there may exist dynamic balance, agents may maximize their central interest through negotiating and finally achieve win-win[2, 6].

2.2 Negotiation Protocol

Negotiation protocol is the precondition and foundation of negotiation, all the interaction must satisfy the rule set and norms. It is defined as follows.

- Agents should have the willing to reach consensus.
- Each period of negotiation should be continuous, and negotiation is carried out by exchanging proposals.
- Agent is concerned with maximizing their own interests, but in the case of that it is difficult to reach an agreement, it should give up some secondary interests while still insist on its central interests.
- Agent shouldn't offer a proposal which is worse than previous one.
- Agent has only three kinds of response to a proposal: accept, then the negotiation is successful; refuse, then the negotiation is failed; counter-proposal, then it will put forward a new proposal by taking into the opposite's proposal and its own state information.
- There is a time limitation. It is decided by state of agents and the issues to be negotiated.

2.3 Negotiation Decision-Making Model

After receiving a proposal offered by other, agent then combines its own state information and the information of current environment, adjusts various dynamic parameters, and finally adopts some methods through its own learning mechanism to determine if the proposal meets its own expectation and then makes a response. This is the negotiation decision-making model of agent, it needs gradually improve, and also it is very important for agent's adaptability and self-governing capacity. To a large extent, the fair or foul of the negotiation decision-making model depends on reference information and agent's learning mechanisms.

3 Multi-agents Competition Model with Negotiation

A good competition model with negotiation should have following attributes: the ability to describe the issue effectively and to support learning ability of agent, the capability to describe the dynamicity and provide a flexible, alternative protocol of competition with negotiation. According to the above several factors, a multi-agent competition model with negotiation, named *MANCM,* is defined as follows.

Definition 1: *MANCM* is a seven tuple $<G,I,O,P,S,R,T>$, where,
G, defined as $\{A_1,A_2,...,A_n\}$,is a set of agents inhabit in the environment, in which A_i is a set of agent of same type. A_i *is defined as* $\{a_1,a_2,...,a_m\}$, in which a_i is a single agent entity.

I, defined as $\{I_1,I_2,...,I_j\}$, is a set of issues that agents need to compete with negotiation, j is the number of topics.

O, defined as $\{O_1,O_2,...,O_k\}$, is a set of competition goal, which represents agent's interests, and it can be a single target or multiple targets.

P, defined as $\{P_1,P_2,...,P_l\}$, is a set of negotiation protocol, in which P_l represents the rules agent should comply with during competition with negotiation.

S is a set of strategy depends on which agents select and carry out their behaviors to interact with other agents, it is defined as $\{S_1,S_2,...,S_r\}$. Agent's strategy set needs to be gradually improved, requires continuous learning and updating, so as to help agent to make more accurate judgments in the following negotiation.

R is a set of the result state of competition with negotiation; it is defined as $\{Accept, Refuse, Propose\}$. *Accept* means agent accepts other's proposals, and reaches an agreement. *Refuse* means that it can not accept the other's proposals and thinks it is not necessary to continue negotiating, so it refuses the proposal and ends negotiation. *Propose* means it doesn't accept the proposal and puts forward its own new proposal.

T is time limitation; different agents have different time limitation. To a certain extent, it will affect the process of competition with negotiation.

3.1 Agent-State Information Model

During the process of competition with negotiation, the information agent own plays a very important role. The more detailed and accurate information agent own, the more correct judgments agent can make, and therefore it will be in a dominant position when compete with others. Each agent has its own state information, which all the selections of strategy are based on [5]. The state information of agent is defined as follows.

Definition 2: *ASIM* is a seven-tuple $<B,S,O,N,E,W,F>$, where,
B is the initial belief of agent, which is expert knowledge and strategy set defined by designers or users.

S, *defined as* $<S_{best},S_{ave},S_{worst}>$, is a set of agent's own historical information, *the elements of* which respectively represents the information of agent's own best, average and the worst behavior. They are respectively endowed with corresponding weight SW_1, $SW2$, $SW3$.

O, *defined as* $<O_{best},O_{ave},O_{worst}>$, represents the historical information of the opponent, *the elements of* which respectively represents the information of opponent's best, average and the worst behavior, Likewise, they are respectively endowed with corresponding weight OW_1,OW_2,OW_3.

N, defined as $<N_{best}, N_{ave}, N_{worst}>$, represents the historical information of other agents in its neighborhood, *the elements of* which respectively represents the information of its own best behavior, the opponent's average behavior and worst behavior, and they are respectively endowed with corresponding weight NW_1, NW_2, NW_3.

E is the information of environment agent inhabit in.

W, defined as $<W_S, W_O, W_N, W_E>$,is the weight set, the elements of which respectively correspond to *S,O,N,E*.

F is the agent's field of vision. *F* determines that agent can only obtain information from other agents within its eye shot.

3.2 Negotiating and Learning Based on Historical Belief

Negotiation strategy is the most significant element in the negotiation, it has great meaning to agent's gain, as well as to the efficiency and success rate of negotiation.

Agent's historical information of negotiation is its main learning material, and agent's learning mechanism has great affect to the improvement of negotiation strategy.

1) Sort out negotiation information: Agent endows different information with corresponding weight according to its referential value, and finally extract the central negotiation information to predict the opponent's strategy as accurately as possible.

INF is the negotiation information finally extracted, $INF = W_S*S + W_O*O + W_N*N + W_E*E$. *S,O,N,E* respectively represents its own history information, the opponent's history information, other neighboring agents' history information and environmental information, where,

$$S = SW_1*S_{best} + SW_2*S_{ave} + SW_3*S_{worst} . \tag{1}$$

$$O = OW_1*O_{best} + OW_2*O_{ave} + OW_3*O_{worst} \tag{2}$$

$$N = NW_1*N_{best} + NW_2*N_{ave} + NW_3*N_{worst} \tag{3}$$

2) Evaluate proposal: When receiving other agent's proposal, it will determines the attitude of the other side through the proposal. If it is uncompromising to refuse to concede and the proposal is far away from its own expectation, it may think that it is unlikely to agree with each other, and then it refuses the proposal, ends the negotiation and starts looking for new partners. If it is cooperative, then agent also adopts a cooperative attitude. If the proposal comes up to agent's expectations, then agent will accept the proposal in a larger probability and end negotiation. Otherwise, if they have not yet reached an agreement, the negotiation will continue and agent puts forward the counter-proposal.

3) Counter-proposal: Agent judges the opponent's intentions according to the proposal and the negotiation information, combines with its own interests and other factors such as current environmental information and time limitation, then makes out the candidate set of negotiation strategy, finally agent selects the best strategy in probability and puts forward the counter-proposal.

4) Learning mechanism of negotiation strategy: After each round of negotiation, agent updates its state information, including agent's own historical information, opponent's historical information, neighboring agents' historical information and environmental information. In order to evaluate the validity and reliability of this information more accurately, finally determine which piece of information has more referential value, we can determine the heavy or light of all the information according to the opponent's proposals after several rounds of negotiation. Thereby agent can forecast more accurately about the strategy the opponent may adopt and increase the weight of corresponding information.

3.3   Flow Chart Of Negotiation Competition Algorithm

The flow chart of negotiation competition algorithm is described as Fig. 1.

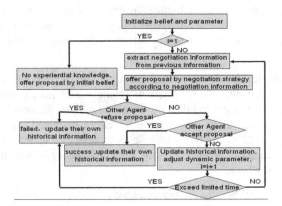

Fig. 1. Flow Chart of Negotiation Competition Algorithm

4   Experiment Simulation

In order to verify the effectiveness of MANCM, this paper constructs an experiment to simulate a competition on price between real estate developers and home buyers. The experiment environment is a two dimension grid, in which there are a certain amount of houses and public facilities which will affect the value of house, such as school, hospital, supermarket, bus station and factory. Value of house is determined by hedonic pricing models [2,4].

4.1   Design of Experiment

According to *MANCM*, the competition between real estate developers and home buyers is designed as follows.

   G includes two kinds of agents, one is *EA* that is a real estate developer community, and the other is *HBA* which denotes a group of home buyers.

HP represents the issue of price; both sides start competing on price by negotiation and ultimately determine a mutually satisfactory price.

O, defined as {OEA, OHBA}, in which OEA represents the goal of EA, and OHBA represents the goal of HBA. The competitive goal of *EA* is maximizing its profit, while the goal of *HBA* is buying a house in the lowest price.

S, defined as {S_{EA},S_{HBA}}, in which S_{EA} is the negotiation strategy of *EA*, S_{HBA} is the negotiation strategy of *HBA.* At the beginning of negotiation, *EA* computes *Value-EA*(value of house computed by *EA*) by considering environmental information around the house, while *HBA* computes *value-HBA*(value of house computed by *HBA*) also by considering environment information around the house. Then they respectively collect the transaction prices of other houses in their neighborhood, the quotation information offered by the opponent and their failure experience of the negotiations, and finally put forward a proposal according to these information and its own history information. Taking account into the influence of time limitation and some other factors, they can make a quoted price to maximize its own interests in probability.

To some extent, maximizing their own interests in a certain probability can promote the success rate of negotiation and reduce the negotiation time.

R, defined as {Accept,Refuse,Propose}. in which *Accept* means agents make a deal and end negotiation. *Refuse* denotes agent refuses quoted price offered by other agents and end negotiation. *Propose* means agent puts forward its own quoted price according to the one it has received.

T is the time limitation. In this experiment, we use rounds of negotiation as time limitation. The number of rounds is determined by Poisson distribution.

4.2 Design of Agent State Information

State information of agent is designed as follow. $EHBASIM = <B_{EHB}, S_{EHB}, O_{EHB}, N_{EHB}, E_{EHB}, W_{EHB}, F_{EHB}>$, where,

B_{EHB} is the agent's belief, including the evaluating mechanism of value of house, the proposal strategy of the initial price and learning mechanisms.

S_{EHB} is history information of *EA*, including the highest bid, the lowest bid and the average bid, and weight of information is adjusted by its influence over the progress of negotiation.

O_{EHB} is information of opponent, including the highest bid, the lowest bid and the average bid of opponent.

N_{EHB} is history information of other agents in the neighborhood, including the best bid of themselves, the average bid of opponents and the worst bid of themselves. Weight of information is adjusted by its influence over the progress of negotiation.

E_{EHB} is the information of environment, including the current proportion between number of houses and number of house-buyers and the average, the highest, the lowest transaction price of houses which had been sold.

F_{EHB} is view field of agent, in this experiment F_{EHB} of *EA* is set as 10, while F_{EHB} of *HBA* is set as 5.

In the state information of agent, the weight of information is adjusted according to the type of agent and the difference between agents' predicting strategy about opponent and the actual one of opponent.

98   L. Wu, C. Yu, and H. Wang

4.3   Experiment Data

Part of the environmental data is given in table 1.

Table 1. Environmental data

Two-dimensional environment	60 * 60
The number of houses	300
The number of developers	45
The number of buyers	800
The number of hospitals	15
The number of supermarkets	25
The number of schools	20
The number of bus stops	30
The number of factories	10

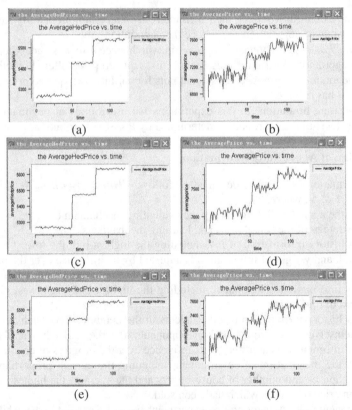

Fig. 2. (a),(c),(e) show the average value of house, while (b),(d),(f) show the average transaction price of house. (a),(b) is the results of competition in which EA doesn't learn about state information and always chooses strategy which can maximize its interest, (c),(d) is the results of competition in which EA learns about state information and always chooses strategy which can maximize its interest, and (e),(f) is the results of competition in which EA doesn't learn about state information and chooses strategy to maximize its interest in probability.

4.4 Analysis of Experiment Results

Based on the simulation platform Swarm and used Java as program language, this paper simulated the competition on house price between buyers and developers with negotiation. In this experiment, in order to better analyze the effect of agents' adaptive learning, the real estate developer agents are design to learn to adapt, while the buyers don't. The experiment results are showed in fig.2.

The experiment results demonstrate that this model can improve the predicted accuracy of opponent's strategy by means of learning negotiation information and successful or failed experimental knowledge of other agents in the neighborhood, raise the profit, shorten negotiation time effectively depends on selecting strategies to maximize its interests in probability, and finally achieve a win-win situation for both parties in competition and improve the success rate and efficiency.

5 Conclusion

Negotiation information is necessary for agent to learn and generate candidate set of competition strategy, selecting the best strategy to maximize its interest in probability also is very helpful to raise the success rate of competing with negotiation and shorten the time. Based on the two factors above, this paper puts forward multi-agents competition model with negotiation. Because of the complex and volatile environment, how to effectively learn negotiation information, more accurately predict opponent's strategy and improve the efficiency of negotiation still is important work in the future.

Acknowledgments. This work was supported in part by a grant from the National Natural Science Foundation of China (No.60805042) and a grant from Program for New Century Excellent Talents in Fujian Province.

References

1. Jennings, N.R.: Automated negotiation. J. Group Decision and Negotiation 10, 199–215 (2001)
2. Feng-qin, C.: Study and Simulation of Artificial Community Negotiation Competition Model. D. Fuzhou University (2010)
3. Qin-ping, Y., Guo-lin, P., Gang, W., Yu-hui, Q.: Research on Mutli-Agent Negotiation Base on Human-computer Interaction. J. Computer Science 35, 226–229 (2008)
4. De, W., Wanshu, H.: Hedonic House Pricing Method and Its Application in Urban Studies. J. City Planning Review 29, 62–70 (2005)
5. Yun-hong, N., Jin-lan, L., De-gan, Z.: Intelligent order online negotiation model with incomplete information of opponent. J. Journal of Computer Applications 29, 221–223 (2009)
6. Juan, W., Yu-mei, C.: Mutli-Agent Negotiation Base on Learning online. D. Zheng Zhou University (2006)

Synchronization of Ghostburster Neurons under External Electrical Stimulation: An Adaptive Approach

Wei Wei[1], Dong Hai Li[2], Jing Wang[3], and Min Zhu[4]

[1] School of Computer and Information Engineering,
Beijing Technology and Business University, Beijing, P.R. China
weiweiustb@yahoo.com.cn
[2] State Key Lab of Power Systems, Department of Thermal Engineering,
Tsinghua University, Beijing, P.R. China
lidongh@mail.tsinghua.edu.cn
[3] Institute of Engineering Research, University of Science and Technology Beijing,
Beijing, P.R. China
wangj@nercar.ustb.edu.cn
[4] Department of Thermal Engineering, Tsinghua University, Beijing, P.R. China
zhumin@tsinghua.edu.cn

Abstract. The synchronization of two Ghostburster neurons under different external electrical stimulations is considered. Firstly, the periodic and chaotic dynamical behaviors of single Ghostburster neuron under various external electrical stimulations are analysed. Then the synchronization of general master-slave chaotic systems is formulated and an adaptive controller based dynamic compensation is designed to synchronize two Ghostburster neurons. Since the adaptive controller based on dynamic compensation is utilized, the exact knowledge of the systems is not necessarily required. Asymptotic synchronization can be achieved by choosing proper controller parameters. Simulation results confirm that the adaptive control approach employed in this paper is valid in the synchronization of two Ghostburster neurons.

Keywords: synchronization, Ghostburster neurons, adaptive control, dynamic compensation.

1 Introduction

The research of nonlinear dynamics in neurons has been attracting a great deal of attention. Efforts in understanding how ensembles of neurons process information have been made by many researchers. Synchronization, a universal phenomenon in natural world [1], also exists and plays a key role in nervous systems [2, 3]. Mammalian nervous systems exhibit various synchronized behaviors, such as periodic, quasi-periodic, chaotic, noise-induced and noise-enhanced synchronous rhythms [4-8]. Synchronization of coupled neurons has been suggested as an important mechanism for accomplishing critical functional goals including biological information processing [9, 10] and the production of regular rhythmical activity [11]. The presence, absence and the degree of synchronization can be a crucial factor of function or dysfunction of a

K. Li et al. (Eds.): LSMS / ICSEE 2010, Part I, CCIS 97, pp. 100–116, 2010.

biological system [12]. Therefore, the synchronization of neural systems has been extensively studied.

To study the nonlinear dynamics and their synchronization of neurons, several nonlinear dynamic models have been presented. The first complete mathematical model describing neural membrane dynamics, i.e. Hodgkin-Huxley (HH) model, was proposed by Hodgkin and Huxley in 1952 [13]. From then on, other neuronal models have been published, such as FitzHugh-Nagumo (FHN) model [14], Hindmarsh-Rose (HR) model [15], Chay model [16], Morris-Lecar neuron model [17], leech neurons [18], Ghostburster neurons [19] and so forth. Meanwhile, a variety of synchronization approaches have been proposed to synchronize the neural systems aforementioned [20-29, 32, 33]. In Ref. [20], exact feedback linearization control is designed to realize the synchronization of two HH neurons. However, to implement the ideal state feedback control law is not practical. Hence, a high-gain observer is employed to modify the ideal control law. The same approach is utilized in the synchronization of FitzHugh-Nagumo neurons under external electrical stimulation [21, 22]. High order slide-mode control [23] and H_∞ variable universe adaptive fuzzy control [24] are also utilized to synchronize the HH neurons. In Ref. [25], a nonlinear controller based on the prior knowledge of the structure and parameters of neural systems is designed to synchronize the coupled FitHugh-Nagumo systems. Internal mode control [26], backstepping control [27], adaptive neural network H_∞ control [28], and impulsive synchronization approach [29] are utilized to realize the synchronization of various neural systems.

In this paper, the synchronization of Ghostburster neurons is considered. Ghostburster neuron model is a two-compartment of pyramidal cell in the electrosensory lateral line lobe (ELL) from weakly electric fish. The model describes the dynamics of soma and dendrite region and the behaviors of membrane potential under different external electrical stimulations [30, 31]. The chaotic and periodic behaviors can be observed by exerting various external stimulations on individual Ghostburster neuron [28]. References [28, 32, 33] discuss the synchronization of Ghostburster neurons under external electrical stimulations by adaptive neural network H_∞ control, H_∞ variable universe fuzzy adaptive control and a model based active control respectively.

As a matter of fact, uncertainties are ubiquitous, and a synchronization approach which is independent of the faithful model of neural systems is of great theoretical and practical value. Here, an adaptive controller based dynamic compensation is adopted to synchronize two Ghostburster neurons under different external stimulations. All of the uncertainties can be observed and compensated by the integral actions of the adaptive controller and the exact synchronization of two unidirectional neurons is achieved.

The rest of this paper is organized as follows. In Section 2, the Ghostburster neuron model and its complex dynamic behaviors are analysed. The synchronization of general master-slave chaotic systems is formulated and an adaptive controller based dynamic compensation is designed to synchronize two Ghostburster neurons in Section 3. Simulation results are given in Section 4 to confirm that the controller is capable of realizing the synchronization of two Ghostburster neurons. Finally, conclusions are drawn in Section 5.

2 Nonlinear Model of Ghostburster Neurons and Its Dynamics

2.1 The Ghostburster Neuron Model

The Ghostburster neuron model identified in pyramidal cells in ELL of weakly electric fish was described in literature [6]. The model consists of two isopotential compartments, representing the soma and the dendrite of the neuron [30]. The two-compartment model of an ELL pyramidal cell is shown in Fig. 1, where one compartment is somatic region, the other is the entire proximal apical dendrite [32] (One can find descriptions on Fig. 1 in Ref. [32] for details). The equations are given in Eqs. (1)-(2).

Soma:

$$\frac{dV_s}{dt} = I_s + g_{Na,s}m_{\infty,s}^2(V_s)(1-n_s)(V_{Na}-V_s) + g_{Dr,s}n_s^2(V_K-V_s) + \frac{g_c}{k}(V_d-V_s) + g_{leak}(V_L-V_s),$$

$$\frac{dn_s}{dt} = \frac{n_{\infty,s}(V_s)-n_s}{\tau_{n.s}}. \tag{1}$$

Dendrite:

$$\frac{dV_d}{dt} = g_{Na,d}m_{\infty,d}^2(V_d)h_d(V_{Na}-V_d) + g_{Dr,d}n_d^2 p_d(V_K-V_d) + \frac{g_c}{1-k}(V_s-V_d) + g_{leak}(V_L-V_d),$$

$$\frac{dh_d}{dt} = \frac{h_{\infty,d}(V_d)-h_d}{\tau_{h.d}},$$

$$\frac{dn_d}{dt} = \frac{n_{\infty,d}(V_d)-n_d}{\tau_{n.d}}, \tag{2}$$

$$\frac{dp_d}{dt} = \frac{p_{\infty,d}(V_d)-p_d}{\tau_{p.d}}.$$

where V_s is somatic membrane potential, V_d is dendritic membrane potential. Table 1 gives out part of the parameter values, each ionic currents ($I_{Na,s}$, $I_{Dr,s}$, $I_{Na,d}$, and $I_{Dr,d}$) is composed of maximal conductance g_{max} (mS/cm^2), infinite conductance curves including both $V_{1/2}$ and k parameters $m_{\infty,s}(V_s) = \dfrac{1}{1+e^{-(V_s-V_{1/2})/k}}$, and time constant τ(ms).

x/y means the channels with both activation (x) and inactivation (y), N/A means the channel activation tracks the membrane potential instantaneously. The rest parameters are chosen to be $k=0.4$, $V_{Na}=40$mV, $V_K=-88.5$mV, $V_{leak}=-70$mV, $g_c=1$, $g_{leak}=0.18$.

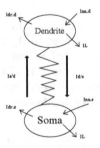

Fig. 1. Scheme of two-compartment model

Table 1. Parameter values for Ghostburster neuron model

Current	$g_{max}(\text{mS/cm}^2)$	$V_{1/2}(\text{mV})$	$\lambda(\text{mV})$	$\tau(\text{ms})$
$I_{Na,s}[n_{\infty,s}(V_s)]$	55	-40	3	0.39
$I_{Dr,s}[m_{\infty,s}(V_s)]$	20	-40	3	N/A
$I_{Na,d}[m_{\infty,d}(V_d)/$ $h_{\infty,d}(V_d)]$	5	-40/-52	5/-5	N/A/1
$I_{Dr,d}[n_{\infty,d}(V_d)/$ $p_{\infty,d}(V_d)]$	15	-40/-65	5/-6	0.9/5

2.2 Dynamic Behaviors of Ghostburster Neuron Model

As external electrical I_s varies, different dynamics can be observed. The plotting of largest Lyapunov exponents varies with I_s can be found in Ref. [32]. In this paper, we choose I_s=6.5mA and I_s=9mA, Ghostburster neuron displays periodic and chaotic behaviors respectively. The initial value of each state variable is chosen to be $[V_{s0}, n_{s0}, V_{d0}, h_{d0}, n_{d0}, p_{d0}]^T=[0,0,0,0,0,0]^T$. Corresponding state response curves of V_s and V_d are shown in Fig. 2.

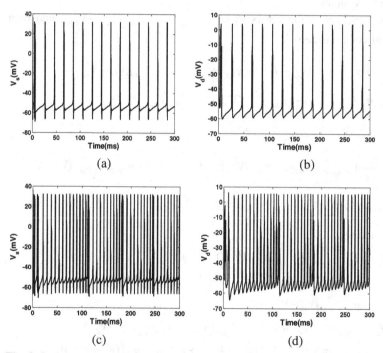

Fig. 2. State response of V_s and V_d when I_s=6.5mA ((a), (b)) and I_s=9mA ((c), (d))

3 Adaptive Synchronization of Ghostburster Neurons

3.1 Synchronization of Two Ghostburster Neurons

In this section, we will consider the synchronization of two Ghostburster neurons by external control action. The master and slave neural systems are shown in Eq. (3).

$$\frac{dV_{s,m}}{dt} = I_{s,m} + g_{Na,s}m_{\infty,s}^2(V_{s,m})(1-n_{s,m})(V_{Na}-V_{s,m}) + g_{Dr,s}n_{s,m}^2(V_K-V_{s,m})$$

$$+\frac{g_c}{k}(V_{d,m}-V_{s,m}) + g_{leak}(V_L-V_{s,m}),$$

$$\frac{dV_{d,m}}{dt} = g_{Na,d}m_{\infty,d}^2(V_{d,m})h_{d,m}(V_{Na}-V_{d,m}) + g_{Dr,d}n_{d,m}^2 p_{d,m}(V_K-V_{d,m}) + \frac{g_c}{1-k}(V_{s,m}-V_{d,m})$$

$$+ g_{leak}(V_L-V_{d,m}),$$

$$\frac{dV_{s,s}}{dt} = I_{s,s} + g_{Na,s}m_{\infty,s}^2(V_{s,s})(1-n_{s,s})(V_{Na}-V_{s,s}) + g_{Dr,s}n_{s,s}^2(V_K-V_{s,s}) \qquad (3)$$

$$+\frac{g_c}{k}(V_{d,s}-V_{s,s}) + g_{leak}(V_L-V_{s,s}) + u_1,$$

$$\frac{dV_{d,s}}{dt} = g_{Na,d}m_{\infty,d}^2(V_{d,s})h_{d,s}(V_{Na}-V_{d,s}) + g_{Dr,d}n_{d,s}^2 p_{d,s}(V_K-V_{d,s}) + \frac{g_c}{1-k}(V_{s,s}-V_{d,s})$$

$$+ g_{leak}(V_L-V_{d,s}) + u_2.$$

u_1 and u_2 are two independent controllers utilized to synchronize the master and the slave neural systems.

Define synchronization errors to be $e_1 = V_{s,s}-V_{s,m}, e_2 = V_{d,s}-V_{d,m}$, and then error dynamical system is derived,

$$\dot{e}_1 = -(\frac{g_c}{k}+g_{leak})e_1 + [(I_{s,s}-I_{s,m}) + g_{Na,s}m_{\infty,s}^2(V_{s,s})(1-n_{s,s})(V_{Na}-V_{s,s})$$

$$+ g_{Dr,s}n_{s,s}^2(V_K-V_{s,s}) - (g_{Na,s}m_{\infty,s}^2(V_{s,m})(1-n_{s,m})(V_{Na}-V_{s,m})$$

$$+ g_{Dr,s}n_{s,m}^2(V_K-V_{s,m}))] + \frac{g_c}{k}e_2 + u_1, \qquad (4)$$

$$\dot{e}_2 = -(\frac{g_c}{1-k}+g_{leak})e_2 + [g_{Na,d}m_{\infty,d}^2(V_{d,s})h_{d,s}(V_{Na}-V_{d,s}) + g_{Dr,d}n_{d,s}^2 p_{d,s}(V_K-V_{d,s})$$

$$- (g_{Na,d}m_{\infty,d}^2(V_{d,m})h_{d,m}(V_{Na}-V_{d,m}) + g_{Dr,d}n_{d,m}^2 p_{d,m}(V_K-V_{d,m}))] + \frac{g_c}{1-k}e_1 + u_2.$$

here, we define

$$f_1 = (I_{s,s}-I_{s,m}) + g_{Na,s}m_{\infty,s}^2(V_{s,s})(1-n_{s,s})(V_{Na}-V_{s,s})$$

$$+ g_{Dr,s}n_{s,s}^2(V_K-V_{s,s}) - (g_{Na,s}m_{\infty,s}^2(V_{s,m})(1-n_{s,m})(V_{Na}-V_{s,m})$$

$$+ g_{Dr,s}n_{s,m}^2(V_K-V_{s,m})), \qquad (5)$$

$$f_2 = g_{Na,d}m_{\infty,d}^2(V_{d,s})h_{d,s}(V_{Na}-V_{d,s}) + g_{Dr,d}n_{d,s}^2 p_{d,s}(V_K-V_{d,s})$$

$$- (g_{Na,d}m_{\infty,d}^2(V_{d,m})h_{d,m}(V_{Na}-V_{d,m}) + g_{Dr,d}n_{d,m}^2 p_{d,m}(V_K-V_{d,m})).$$

thus error dynamical system (4) can be written as

$$\dot{e}_1 = -(\frac{g_c}{k}+g_{leak})e_1 + \frac{g_c}{k}e_2 + f_1 + u_1,$$

$$\dot{e}_2 = -(\frac{g_c}{1-k}+g_{leak})e_2 + \frac{g_c}{1-k}e_1 + f_2 + u_2. \qquad (6)$$

Our work is to design controller u_1 and u_2, which make the states of slave neural system track those of master neural system, i.e. $\lim_{t\to\infty} e_i(t) = 0$ with the help of controllers u_i, $i=1, 2$.

3.2 Adaptive Controller Design

For convenience, Eq. (3) can be written as following form

$$\begin{aligned}
\dot{x}_m &= Ax_m + Bf(x_m), \\
\dot{x}_s &= Ax_s + B(f(x_s) + u(t)).
\end{aligned} \tag{7}$$

where $x_m, x_s \in R^n$, $f : R^n \to R$ are bounded smooth nonlinear function, $u(t) \in R$ is the control input,

$$A = \begin{bmatrix} 0 & 1 & 0 & \cdots & 0 \\ 0 & 0 & 1 & \cdots & 0 \\ \vdots & \vdots & \vdots & \vdots & \vdots \\ 0 & 0 & \cdots & 0 & 1 \\ 0 & 0 & \cdots & 0 & 0 \end{bmatrix}_{n\times n}, B = \begin{bmatrix} 0 \\ 0 \\ 0 \\ 0 \\ 1 \end{bmatrix}_{n\times 1}$$

Define $e = x_s - x_m$, the error dynamical system can be written as

$$\dot{e} = Ae + B(f(x_s, x_m) + u(t)). \tag{8}$$

Synchronization of the master and slave systems is to make the error e asymptotically converge to zero by the controllers. Here, an adaptive controller based on the dynamic compensation mechanism (proposed by Tornambe, see Ref. [34] for details) is adopted. The control law has the form

$$\begin{aligned}
u &= -h_0 z_1 - h_1 z_2 - \cdots - h_{n-1} z_n - \hat{d} = -\sum_{i=0}^{n-1} h_i z_{i+1} - \hat{d}, \\
\hat{d} &= \xi + \sum_{i=0}^{n-1} k_i z_{i+1}, \\
\dot{\xi} &= -k_{n-1}\xi - k_{n-1}\sum_{i=0}^{n-1} k_i z_{i+1} - \sum_{i=0}^{n-2} k_i z_{i+2} - k_{n-1} u.
\end{aligned} \tag{9}$$

Parameters h_i, $i=0,1,\ldots,n-1$, are chosen so that the eigenvalues of polynomial $h(s) = s^n + h_{n-1}s^{n-1} + \cdots + h_1 s + h_0$ are in the open left half-plane.
Substituting Eq. (9) into dynamical error system (8), we have

$$\dot{e}_n = -\sum_{i=0}^{n-1} h_i z_{i+1} + (f(x_s, x_m) - \hat{d}). \tag{10}$$

$\tilde{d} \triangleq f(x_s, x_m) - \hat{d}$, Eq. (10) can be written as

$$\dot{e}_n = -\sum_{i=0}^{n-1} h_i z_{i+1} + \tilde{d}. \tag{11}$$

Therefore, the whole error dynamical system is

$$\dot{e} = Fe + q\tilde{d}. \tag{12}$$

where $F = \begin{bmatrix} 0 & 1 & 0 & \cdots & 0 \\ 0 & 0 & 1 & \cdots & 0 \\ \vdots & \vdots & \vdots & \vdots & \vdots \\ 0 & 0 & \cdots & 0 & 1 \\ -h_0 & -h_1 & \cdots & \cdots & -h_{n-1} \end{bmatrix}_{n \times n}$, $q = \begin{bmatrix} 0 \\ 0 \\ 0 \\ 0 \\ 1 \end{bmatrix}_{n \times 1}$

Theorem [34]. There exists a constant value $\mu^* > 0$, such that if $k_{n-1} > \mu^*$, then the closed-loop system (8) and (9) is asymptotically stable.

For the synchronization of Ghostburster neurons, we have following corollary.

Corollary. There exist two constants $\mu_1^* > 0$ and $\mu_2^* > 0$, such that if $k_{01} > \mu_1^*$ and $k_{02} > \mu_2^*$, then closed-loop error dynamical systems (15) and (16) are asymptotically stable.

Proof

For error dynamical system (6), the adaptive control law for u_1 and u_2 can be represented as follows,

$$\begin{aligned}
u_1 &= -h_{01}e_1 - \hat{d}_1, \\
\dot{\xi}_1 &= -k_{01}\xi_1 - k_{01}^2 e_1 - k_{01}u_1, \\
\hat{d}_1 &= \xi_1 + k_{01}e_1, \\
u_2 &= -h_{02}e_2 - \hat{d}_2, \\
\dot{\xi}_1 &= -k_{02}\xi_2 - k_{02}^2 e_2 - k_{02}u_2, \\
\hat{d}_2 &= \xi_2 + k_{02}e_2.
\end{aligned} \tag{13}$$

Substituting Eq. (13) into error dynamical system (6), we have

$$\begin{aligned}
\dot{e}_1 &= G_1(e_1, e_2) - h_{01}e_1 - \hat{d}_1 \triangleq -h_{01}e_1 + d_1 - \hat{d}_1, \\
\dot{e}_2 &= G_2(e_1, e_2) - h_{02}e_2 - \hat{d}_2 \triangleq -h_{02}e_2 + d_2 - \hat{d}_2.
\end{aligned} \tag{14}$$

where $G_1(e_1, e_2) = -(\dfrac{g_c}{k} + g_{leak})e_1 + \dfrac{g_c}{k}e_2 + f_1$,

$G_2(e_1, e_2) = -(\dfrac{g_c}{1-k} + g_{leak})e_2 + \dfrac{g_c}{1-k}e_1 + f_2$.

Let $\tilde{d} = d - \hat{d}$, the derivative of d and $\hat{d}$ are

$$\dot{d}_1 = \frac{d}{dt}(G_1(e_1, e_2)) = a_1(e_1, e_2), \dot{d}_2 = \frac{d}{dt}(G_2(e_1, e_2)) = a_2(e_1, e_2)$$

$$\dot{\hat{d}}_1 = -k_{01}\xi_1 - k_{01}^2 e_1 - k_{01}u_1 + k_{01}(d_1 + u_1) = k_{01}(d_1 - \xi_1 - k_{01}e_1) = k_{01}(d_1 - \hat{d}_1) = k_{01}\tilde{d}_1$$

Similarily, $\dot{\tilde{d}}_2 = k_{02}\tilde{d}_2$. Therefore the two closed-loop systems are

$$\dot{e}_1 = -h_{01}e_1 + \tilde{d}_1 ,$$
$$\dot{\tilde{d}}_1 = a_1(e_1,e_2) - k_{01}\tilde{d}_1 . \tag{15}$$

$$\dot{e}_2 = -h_{02}e_2 + \tilde{d}_2 ,$$
$$\dot{\tilde{d}}_2 = a_2(e_1,e_2) - k_{02}\tilde{d}_2 . \tag{16}$$

Supposing that V_1 and V_2 are differentiable positive definite function of e_1 and e_2, i.e.

$$V_1 = \frac{1}{2}e_1^2 , \ V_2 = \frac{1}{2}e_2^2$$

Then the compact domain $U_{V_1,M_1} = \{e_1 \mid V_1 = \frac{1}{2}e_1^2 \le M_1\}$ and $U_{V_2,M_2} = \{$

$\{e_2 \mid V_2 = \frac{1}{2}e_2^2 \le M_2\}$ can be defined. M_1 and M_2 are arbitrary positive constants. Thus,

for any $h_{01} > 0, h_{02} > 0$ and any $e_1 \in U_{V_1,M_1}$ and $e_2 \in U_{V_2,M_2}$, there are

1) $V_1(0) = 0, \dfrac{dV_1}{de_1}\big|_{e_1=0} = 0, V_2(0) = 0, \dfrac{dV_2}{de_2}\big|_{e_2=0} = 0$,

2) $\dfrac{dV_1}{de_1}(-h_{01}e_1) \le -e_1^2, \dfrac{dV_2}{de_2}(-h_{02}e_2) \le -e_2^2$.

Stabilize closed-loop systems (15) and (16), two Lyapunov functions are defined

$$W_1 = V_1 + \frac{1}{2}\tilde{d}_1^2 , W_2 = V_2 + \frac{1}{2}\tilde{d}_2^2 .$$

For fixed M_1 and M_2, let $U_{W_1,M_1} , U_{W_2,M_2}$ be compact domains such that

$$U_{W_1,M_1} = \{(e_1,\tilde{d}_1) \mid W_1 = V_1 + \frac{1}{2}\tilde{d}_1^2 \le M_1\}, \ U_{W_2,M_2} = \{(e_2,\tilde{d}_2) \mid W_2 = V_2 + \frac{1}{2}\tilde{d}_2^2 \le M_2\}$$

Differentiating W_1 and W_2 along systems (15) and (16)

$$\dot{W}_1 = \dot{V}_1 + \tilde{d}_1\dot{\tilde{d}}_1 = \frac{dV_1}{de_1}(-h_{01}e_1 + \tilde{d}_1) + \tilde{d}_1(a_1(e_1,e_2) - k_{01}\tilde{d}_1)$$

$$\dot{W}_2 = \dot{V}_2 + \tilde{d}_2\dot{\tilde{d}}_2 = \frac{dV_2}{de_2}(-h_{02}e_2 + \tilde{d}_2) + \tilde{d}_2(a_2(e_1,e_2) - k_{02}\tilde{d}_2)$$

It is obvious that the projections of U_{W_1,M_1} and U_{W_2,M_2} into hyperplane $\tilde{d} = 0$ coincides with U_{V_1,M_1} and U_{V_2,M_2}. Thus 1) and 2) also hold for $\forall (e,\tilde{d}) \in U_{W,M}$. According to Ref. [34], we have following technical lemmas

a) since $\dfrac{dV}{de}\big|_{e=0} = 0$, then $\left|\dfrac{dV}{de}\right| \le P_V |e| \qquad \forall e \in U_{W,M}\big|_e$

b) $|a(e_1,e_2)| \le P_a |e| \qquad \forall e \in U_{W,M}\big|_e$

where P_V, P_α are positive and dependent on M. Therefore, for any $\forall (e, \tilde{d}) \in U_{W,M}$, we have

$$\dot{W}_1 = \dot{V}_1 + \tilde{d}_1 \dot{\tilde{d}}_1 = \frac{dV_1}{de_1}(-h_{01}e_1 + \tilde{d}_1) + \tilde{d}_1(a_1(e_1, e_2) - k_{01}\tilde{d}_1)$$

$$\leq -e_1^2 + (P_{V_1} + P_{\alpha 1})|e_1||\tilde{d}_1| - k_{01}\tilde{d}_1^2 = -(e_1^2 - (P_{V_1} + P_{\alpha 1})|e_1||\tilde{d}_1| + k_{01}\tilde{d}_1^2)$$

$$= -\begin{bmatrix} |e_1| & |\tilde{d}_1| \end{bmatrix}\begin{bmatrix} 1 & \dfrac{-(P_{V_1} + P_{\alpha 1})}{2} \\ \dfrac{-(P_{V_1} + P_{\alpha 1})}{2} & k_{01} \end{bmatrix}\begin{bmatrix} |e_1| \\ |\tilde{d}_1| \end{bmatrix}$$

$$\dot{W}_2 = \dot{V}_2 + \tilde{d}_2 \dot{\tilde{d}}_2 = \frac{dV_2}{de_2}(-h_{02}e_2 + \tilde{d}_2) + \tilde{d}_2(a_2(e_1, e_2) - k_{02}\tilde{d}_2)$$

$$\leq -e_2^2 + (P_{V_2} + P_{\alpha 2})|e_2||\tilde{d}_2| - k_{02}\tilde{d}_2^2 = -(e_2^2 - (P_{V_2} + P_{\alpha 2})|e_2||\tilde{d}_2| + k_{02}\tilde{d}_2^2)$$

$$= -\begin{bmatrix} |e_2| & |\tilde{d}_2| \end{bmatrix}\begin{bmatrix} 1 & \dfrac{-(P_{V_2} + P_{\alpha 2})}{2} \\ \dfrac{-(P_{V_2} + P_{\alpha 2})}{2} & k_{02} \end{bmatrix}\begin{bmatrix} |e_2| \\ |\tilde{d}_2| \end{bmatrix}$$

According to Sylvester criterion, it is straightforward to obtain following inequalities

$$k_{01} > \frac{(P_{V_1} + P_{\alpha 1})^2}{2}, k_{02} > \frac{(P_{V_2} + P_{\alpha 2})^2}{4}.$$

Let $\mu_1^* = \dfrac{(P_{V_1} + P_{\alpha 1})^2}{2}, \mu_2^* = \dfrac{(P_{V_2} + P_{\alpha 2})^2}{4}$, thus, if $k_{01} > \mu_1^*, k_{02} > \mu_2^*$, $\dot{W}_1, \dot{W}_2$ are negative definite in U_{W_1, M_1} and U_{W_2, M_2}, closed-loop systems (15) and (16) are asymptotically stable in $U_{W,M}$.   □

Closed-loop error dynamical system is asymptotically stable, which means the synchronization of master and slave neural systems is achieved.

4 Numerical Simulations

In this section, numerical simulations are presented to verify the effectiveness and feasibility of the adaptive controller utilized in this paper. The initial states are chosen to be $[V_{s,m0}, n_{s,m0}, V_{d,m0}, h_{d,m0}, n_{d,m0}, p_{d,m0}]^T = [1,0,1,0,0,0]^T$, and $[V_{s,s0}, n_{s,s0}, V_{d,s0}, h_{d,s0}, n_{d,s0}, p_{d,s0}]^T = [0,0,0,0,0,0]^T$. Two group numerical simulations are accomplished. In following simulations, control actions are added at 100ms, the adaptive controller parameters of u_1 and u_2 are chosen to be $k_{01} = k_{02} = 116, h_{01} = h_{02} = 26$.

Group I. External stimulus of the master neuron is chosen to be 6.5mA, and the stimulus of the slave neuron is 9mA. We aim to drive the slave neuron system, which exhibits chaotic behaviors, to track periodic behaviors of master neuron system.

i) Case 1: No system parameters are changed. The time traces of system states, synchronization errors and phase plane diagram after control actions being applied are shown in Fig. 3.

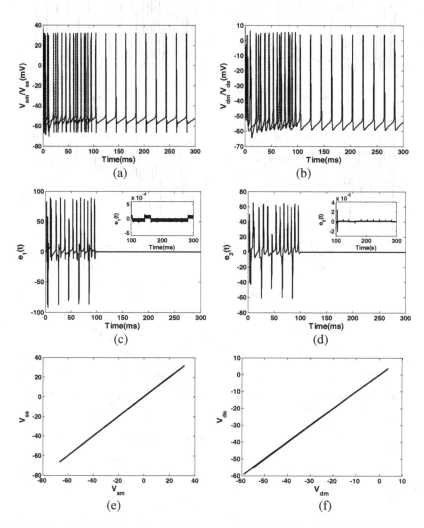

Fig. 3. Synchronization of two Ghostburster neurons via adaptive control. (a) Time traces of system states V_{sm}, V_{ss}. (b) Time traces of system states V_{dm}, V_{ds}. (c) Synchronization errors between V_{sm} and V_{ss}. (d) Synchronization errors between V_{dm} and V_{ds}. (e) Phase plane diagram of V_{sm} and V_{ss} after control actions are activated. (f) Phase plane diagram of V_{dm} and V_{ds} after control actions are activated.

ii) Case 2: g_c and g_{leak} in slave neural system are abruptly changed at 150ms. With a difference of 10% from the nominal values, g_c and g_{leak} are abruptly changed to be 0.9 and 0.162 respectively. Corresponding dynamical responses are given in Fig. 4.

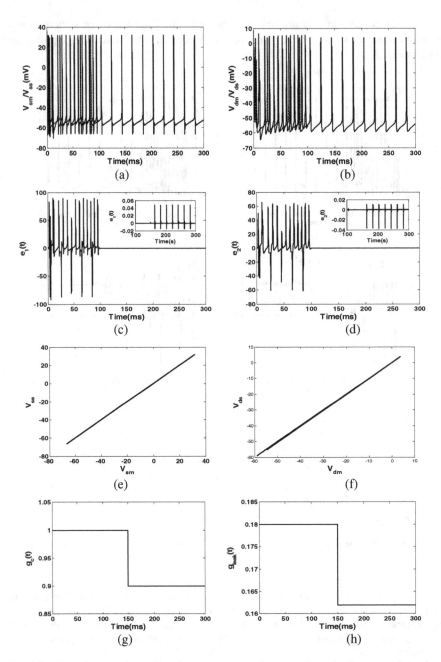

Fig. 4. Synchronization of two Ghostburster neurons when g_c and g_{leak} in slave neural system are abruptly changed at 150ms. (a) Time traces of system states V_{sm}, V_{ss}. (b) Time traces of system states V_{dm}, V_{ds}. (c) Synchronization errors between V_{sm} and V_{ss}. (d) Synchronization errors between V_{dm} and V_{ds}. (e) Phase plane diagram of V_{sm} and V_{ss} after control actions are activated. (f) Phase plane diagram of V_{dm} and V_{ds} after control actions are activated. (g) Variation of g_c. (h) Variation of g_{leak}.

iii) Case 3: Random noises are introduced into g_c and g_{leak} in slave neural system at 150ms. The intensity of the noise is 0.1. The responses are presented in Fig. 5.

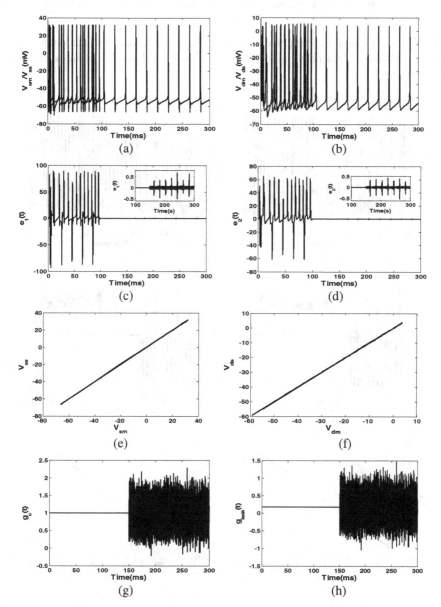

Fig. 5. Synchronization of two Ghostburster neurons when random noises are introduced into g_c and g_{leak} in slave neural system at 150ms. (a) Time traces of system states V_{sm}, V_{ss}. (b) Time traces of system states V_{dm}, V_{ds}. (c) Synchronization errors between V_{sm} and V_{ss}. (d) Synchronization errors between V_{dm} and V_{ds}. (e) Phase plane diagram of V_{sm} and V_{ss} after control actions are activated. (f) Phase plane diagram of V_{dm} and V_{ds} after control actions are activated. (g) Variation of g_c. (h) Variation of g_{leak}.

Group II. External stimulus of the master neuron is chosen to be 9mA, and the stimulus of the slave neuron is 6.5mA. We aim to drive the slave neuron system, which exhibits periodic behaviors, to track chaotic behaviors of master neuron system. The same cases are considered in Group II.

i) Case 1: No system parameters are changed. The time traces of system states, synchronization errors and phase plane diagram after control actions being applied are shown in Fig. 6.

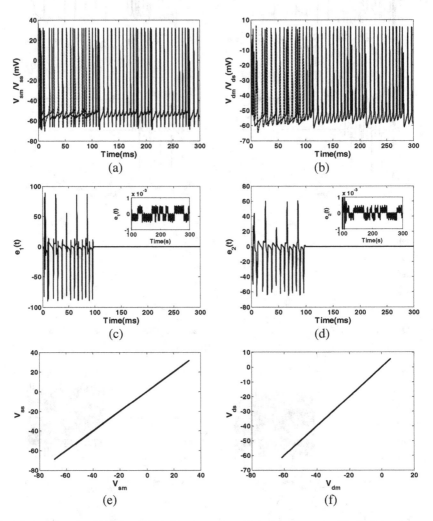

Fig. 6. Synchronization of two Ghostburster neurons via adaptive control. (a) Time traces of system states V_{sm}, V_{ss}. (b) Time traces of system states V_{dm}, V_{ds}. (c) Synchronization errors between V_{sm} and V_{ss}. (d) Synchronization errors between V_{dm} and V_{ds}. (e) Phase plane diagram of V_{sm} and V_{ss} after control actions are activated. (f) Phase plane diagram of V_{dm} and V_{ds} after control actions are activated.

ii) Case 2: g_c and g_{leak} in slave neural system are abruptly changed at 150ms. With a difference of 10% from the nominal values, g_c and g_{leak} are abruptly changed to be 0.9 and 0.162 respectively. (As Group I case 2.) Corresponding dynamical responses are given in Fig. 7.

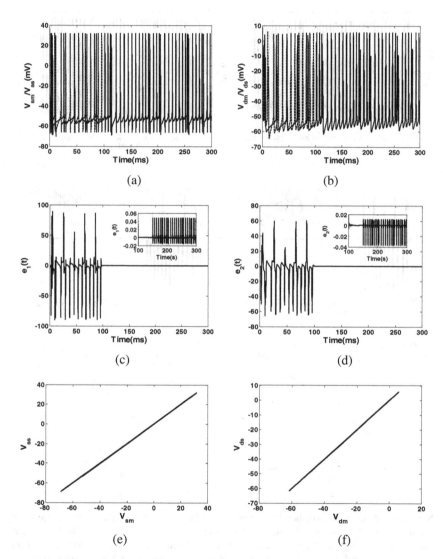

Fig. 7. Synchronization of two Ghostburster neurons when g_c and g_{leak} in slave neural system are abruptly changed at 150ms. (a) Time traces of system states V_{sm}, V_{ss}. (b) Time traces of system states V_{dm}, V_{ds}. (c) Synchronization errors between V_{sm} and V_{ss}. (d) Synchronization errors between V_{dm} and V_{ds}. (e) Phase plane diagram of V_{sm} and V_{ss} after control actions are activated. (f) Phase plane diagram of V_{dm} and V_{ds} after control actions are activated.

iii) Case 3: Random noises are introduced into g_c and g_{leak} in slave neural system at 150ms. The intensity of the noise is 0.1. (As Group I case 3.) The responses are presented in Fig. 8.

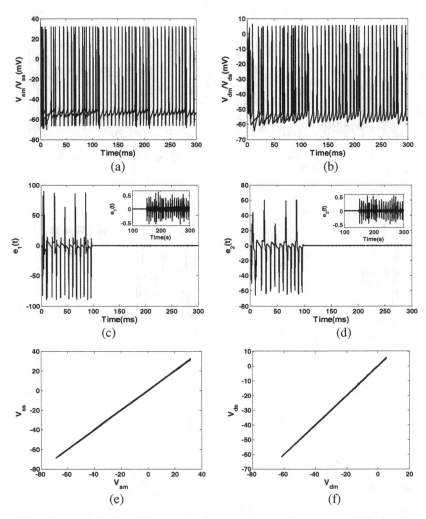

(a)

(b)

(c)

(d)

(e)

(f)

Fig. 8. Synchronization of two Ghostburster neurons when random noises are introduced into g_c and g_{leak} in slave neural system at 150ms. (a) Time traces of system states V_{sm}, V_{ss}. (b) Time traces of system states V_{dm}, V_{ds}. (c) Synchronization errors between V_{sm} and V_{ss}. (d) Synchronization errors between V_{dm} and V_{ds}. (e) Phase plane diagram of V_{sm} and V_{ss} after control actions are activated. (f) Phase plane diagram of V_{dm} and V_{ds} after control actions are activated.

From the simulation results of Group I and II, we can see clearly that the adaptive controller can realize the synchronization of two Ghostburster neurons successfully, no matter whether the system parameters change or not. The simulation results show that the controller has good performance robustness.

5 Conclusions

In this paper, the synchronization of unidirectional Ghostburster neurons is investigated. Both of the theoretical and simulation results show that the adaptive controller can achieve the synchronization of Ghostburster neurons. The perfect knowledge of the neural systems is not required. By choosing proper controller parameters, synchronization can be realized. This is of great importance in practice, because uncertainties are ubiquitous, it is not practical to get exact models of systems.

It is worth pointing out that although the synchronization of unidirectional Ghostburster neurons is considered in this paper, the adaptive controller adopted is model free, which indicates that it can be generalized in the synchronization of other neural systems such as the synchronization of HH neurons, HR neurons, FHN neurons etc.

Acknowledgments

This work was supported by Basic Research Program of China (2007CB210106) and National Key Technology Research and Design Program of China (2006BAA03B04).

References

1. Pikovsky, A., Rosenblum, M., Kurths, J.: Synchronization: A universal Concept in Nonlinear Sciences. Cambridge University Press, New York (2001)
2. Womelsdorf, T., Fries, P.: The role of neuronal synchronization in selective attention. Curr. Opin. Neurobiol. 17(2), 154–160 (2007)
3. Gray, C.M.: Synchronous oscillations in neuronal systems: Mechanisms and functions. J. Comput. Neurosci. 1, 11–38 (1994)
4. Basar, E.: Brain Function and Oscillations I: Brain oscillations, Principles and Approaches. Springer, Berlin (1998)
5. Haken, H.: Branin Dynamics-Synchronization and Activity Patterns in Pulse-Coupled Neural Nets with Delays and Noise. Springer, Berlin (2002)
6. Gray, C.M., König, P., Engel, A.K., Singer, W.: Oscillatory responses in cat visual cortex exhibit inter-columnar synchronization which reflects global stimulus properties. Nature 338, 334–337 (1989)
7. Gray, C.M., Mcormick, D.A.: Chattering cells: Superficial pyramidal neurons contributing to the generation of synchronization oscillations in the visual cortex. Science 274, 109–113 (1996)
8. Wang, Q.Y., Duan, Z.S., Feng, Z.S., Chen, G.R., Lu, Q.S.: Synchronization transition in gap-junction-coupled leech neurons. Physica A 387, 4404–4410 (2008)
9. Meister, M., Wong, R.O., Baylor, D.A., Shatz, C.J.: Synchronous bursts of action potentials in ganglion cells of the developing mammalian retina. Science 252, 939–943 (1991)
10. Kreiter, A.K., Singer, W.: Stimulus-dependent synchronization of neuronal responses in the visual cortex of the awake macaque monkey. J. Neurosci. 16, 2381–2396 (1996)
11. Che, Y.Q., Wang, J., Tsang, K.M., Chan, W.L.: Unidirectional synchronization for Hindmarsh-Rose neurons via robust adaptive sliding mode control. Nonlinear Analysis: Real World Applications 11, 1096–1104 (2010)
12. Shuai, J.W., Durand, D.M.: Phase synchronization in two coupled chaotic neurons. Phys. Lett. A 264, 289–297 (1999)
13. Hodgkin, L., Huxley, A.F.: A quantitative description of membrane and its application to conduction and excitation in nerve. J. Physiol. 117, 500–544 (1952)

14. FitzhHugh, R.: Trashholds and plateaus in the Hodgkin-Huxley nerve equations. J. Gen. Physiol. 43, 867–896 (1960)
15. Hindmarsh, J.L., Rose, R.M.: A model of neuronal busting using three coupled first order differential equations. Proc. Roy. Soc. Lond. B Biol. 221, 87–102 (1984)
16. Chay, T.R.: Chaos in a three-variable model of an excitable cell. Physica D 16, 233–242 (1985)
17. Morris, C., Lecar, H.: Voltage oscillations in the barnacle giant muscle fiber. Biophys. J. 35, 193–213 (1981)
18. Shilnikov, A., Calabrese, R.L., Cymbalyuk, G.: Mechanism of bistability: Tonic spiking and bursting in a neuron model. Phys. Rev. E 71 (2005), 056214-9
19. Doiron, B., Laing, C., Longtin, A.: Ghostbursting : a novel neuronal burst mechanism. J. Comput. Neurosci. 12, 5–25 (2002)
20. Cornejo-Pérez, O., Femat, R.: Unidirectional synchronization of Hodgkin-Huxley neurons. Chaos, Solitons and Fractals 25, 43–53 (2005)
21. Wang, J., Zhang, T., Deng, B.: Synchronization of FitzHugh-Nagumo neurons in external electrical stimulation via nonlinear control. Chaos, Solitons and Fractals 31, 30–38 (2007)
22. Zhang, T., Wang, J., Fei, X.Y., Deng, B.: Synchronization of coupled FitzHugh-Nagumo systems via MIMO feedback linearization control. Chaos, Solitons and Fractals 33, 194–202 (2007)
23. Aguilar-López, R., Martínez-Guerra, R.: Synchronization of a coupled Hodgkin-Huxley neurons via high order sliding-mode feedback. Chaos, Solitons and Fractals 37, 539–546 (2008)
24. Wang, J., Che, Y.Q., Zhou, S.S., Deng, B.: Unidirectional synchronization of Hodgkin-Huxley neurons exposed to ELF electric field. Chaos, Solitons and Fractals 39, 1335–1345 (2009)
25. Ahmet, U., Lonngren, K.E., Bai, E.W.: Synchronization of the coupled FitzHugh-Nagumo systems. Chaos, Solitons and Fractals 20, 1085–1090 (2004)
26. Wei, X.L., Wang, J., Deng, B.: Introducing internal model to robust output synchronization of FitzHugh-Nagumo neurons in external electrical sitmulation. Commu. Nonlinear Sci. Numer. Simulat. 14, 3108–3119 (2009)
27. Deng, B., Wang, J., Fei, X.Y.: Synchronization two coupled chaotic neurons in external electrical stimulation using backstepping control. Chaos, Solitons and Fractals 29, 182–189 (2006)
28. Li, H.Y., Wong, Y.K., Chan, W.L., Tsang, K.M.: Synchronization of Ghostburster neurons under external electrical stimulation via adaptive neural network H_∞ control. Neurocomputing (2010), doi:10.1016/j.neucom.2010.03.004
29. Wu, Q.J., Zhou, J., Xiang, L., Liu, Z.R.: Impulsive control and synchronization of chaotic Hindmarsh-Rose models for neuronal activity. Chaos, Solitons and Fractals 41, 2706–2715 (2009)
30. Laing, C.R., Doiron, B., Longtin, A., Maler, L.: Ghostbursting: the effects of dendrites on spike patterns. Neurocomputing 44-46, 127–132 (2002)
31. Oswald, A.M., Chacron, M.J., Doiron, B., et al.: Parallel processing of sensory input by bursts and isolated spikes. The Journal of Neuroscience 24(18), 4351–4362 (2004)
32. Wang, J., Chen, L., Deng, B.: Synchronization of Ghostburster neuron in external electrical stimulation via H_∞ variable universe fuzzy adaptive control. Chaos, Solitons and Fractals 39, 2076–2085 (2009)
33. Sun, L., Wang, J., Deng, B.: Global synchronization of two Ghostburster neurons via active control. Chaos, Solitons and Fractals 40, 1213–1220 (2009)
34. Tornambé, A., Valigi, P.: A decentralized controller for the robust stabilization of a class of MIMO dynamical systems. Journal of Dynamic Systems, Measurement and Control 116, 293–304 (1994)

A Collision Detection System for an Assistive Robotic Manipulator

Weidong Chen[1,2], Yixiang Sun[1,2], and Yuntian Huang[1,2]

[1] Department of Automation, Shanghai Jiao Tong University, Shanghai 200240, China
[2] State Key Laboratory of Robotics and System (HIT), Harbin 150001, China
wdchen@sjtu.edu.cn

Abstract. To make human-manipulator interaction safe, a method and its reali-zation of safety design of assistive robotic manipulator based on collision detec-tion is presented in this paper. The collision is detected by the difference of the reference torque calculated according to the dynamic model and the factual tor-que measured by torque sensors. In the design of the joint torque sensor, the fi-nite element analysis is applied to determine the optimal position for pasting strain gauge, and then a signal processing circuit with high capacity of resisting disturbances is developed. According to the low speed characteristic of the as-sistive robotic manipulator, a simplified dynamic model is established, which balances the efficiency and accuracy of the calculation of the reference torque. Experimental results are given to prove the validity of the proposed design.

Keywords: Assistive Robotic Manipulator; Safety; Collision Detection; Torque Sensor.

1 Introduction

Assistive robot manipulators can help the old or the upper-limb-disabled people with their daily life by providing services such as fetching objects, opening/closing doors, food intake so as to improve their life quality. Since an assistive robotic manipulator's working area is very close to people, measures must be taken to guarantee people's safety, which means special requirements like real-time collision detection is needed.

Researches focusing on the safety techniques are divided into two classes: one is to assure the safety in mechanical design phase, the other is to improve the algorithm in system control phase. As for the second class, safety is supposed to be improved via a certain control system including ACC(Active Compliance Control)[3], driver torque control[4] etc. The key to these algorithms is instant and accurate collision detection.

Collision detection techniques applied to assistive robotic manipulators mainly base on a 6 degree F/T sensor mounted on the end actuator, skin-shaped tactile sen-sors to detect the external force [5]. Disadvantages of the above methods are: using a 6 degree F/T sensor can only detect the external force act on the end actuator. The skin-shaped tactile sensors mounted all over the manipulator give no blind area but the wiring is complex and reduces the hardware robustness.

K. Li et al. (Eds.): LSMS / ICSEE 2010, Part I, CCIS 97, pp. 117–123, 2010.

A collision detection method based on joint torque sensors is introduced in this paper: three torque sensors are mounted on three joints of the manipulator to detect joints' actual output torques. By comparing the actual torque with the reference value calculated from the manipulator's simplified dynamic model, an effective and easy to install collision detection system is realized.

2   Structure of Collision Detection System

Fig.1 illustrates the block diagram of the collision detection system for a single joint. Three torque sensors are mounted on three joints, they detect the output torque(τ_i) of the joints. Joint position($\mathbf{q}$) and velocity($\dot{\mathbf{q}}$) are obtained from joint drivers and by differentiating the velocity with respect to time can we obtain the acceleration($\ddot{\mathbf{q}}$). Utilizing the simplified dynamic model of the manipulator, reference output torque under no external force(τ_r) can be calculated, thus the error($\Delta\tau$) between the actual and reference torque are obtained. Set proper threshold(γ_i) for each torque sensor so that when $|\Delta\tau_i|$ gets larger than γ_i, a collision's considered to have happened.

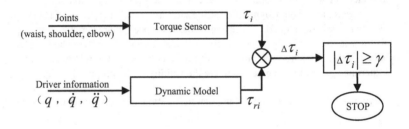

Fig. 1. Structure of collision detection system

3   Design of Joint Torque Sensors

3.1   Structure Design of Joint Torque Sensors

Three torque sensors are respectively mounted on the manipulator's three joints' transmission shafts so as to detect the output torques of the joints. Each sensor contains two parts: a torque disk and a signal processing circuit. Strain foils on one torque disk are connected to form an electric bridge so that shear deformation of the disk can generate a voltage variation between the bridge's two ends. After amplifying, filtering and calibration[6], the signal is enlarged enough for use.

One torque disk has 8 cross beams[7], which provides a linear relation between the electric signal and the joint's output torque. Because the LY2 material has the merits like high proportional limit, light-weight, nice processing property, high rigidity and natural vibration frequency, it is used to make the torque disks that are sensitive in torque detection while the output signal stays steady and has high repetitiveness.

The electric resistance of each strain foil is 350 Ω and they are affixed to 4 out of 8 beams. Finite element analysis was carried out to find the places on the disk where deformation caused by shear force happens the greatest and that caused by other forces happens the least. Fig.2 shows the schematic drawing and a photo of a real torque disk to which strain foils are affixed. The strain foils are connected to form an electric bridge whose structure can effectively suppress the influences caused by temperature drift and cross force acting on the torque disk[8].

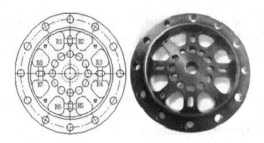

Fig. 2. Joint torque sensor

3.2 Signal Processing Circuit Design

The signal processing circuit includes power, amplifier and filter modules. 24V is transformed to different voltages by the power module to meet the needs of the rest two modules. The signal generated by the electric bridge is trivial so that a high-precision amplifier circuit is designed by using an instrumentation amplifier which has low temperature drift and can suppress the common-mode interference.

The amplified signal is filtered by an active second-order low-pass filter whose cut-off frequency is set to 1 kHz so as to smooth the high frequency noise. Fig.3 illustrates the block diagram of the signal processing circuit.

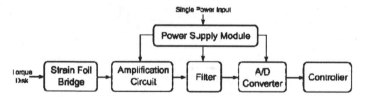

Fig. 3. Diagram of signal processing circuit

4 Dynamic Model of Assistive Robotic Manipulator

4.1 Dynamic Model Building

The dynamic model of the robotic manipulator can be described as follows[9]:

$$\mathbf{B(q)\ddot{q} + C(q,\dot{q}) + g(q) = \tau_r - J^T(q)h} \tag{1}$$

where $\mathbf{q, \dot{q}, \ddot{q}}$ notes for the position, velocity, acceleration, $\mathbf{h}$ notes for the external force acting on the end actuator, $\mathbf{g(q)}$ notes for gravity force.

Rearrange eq.1 we have the expression for the reference torque $\boldsymbol{\tau}$:

$$\boldsymbol{\tau}_r = \mathbf{B(q)\ddot{q}} + \mathbf{C(q, \dot{q})} + \mathbf{g(q)} + \mathbf{J^T(q)h} \qquad (2)$$

4.2 Simplification of the Dynamic Model

Assistive robotic manipulators are light-weighted; have small inertia, joint velocity and acceleration. So in practice, the influence of velocity and acceleration to the manipulator can be neglected, thus the dynamic model can be approximated to:

$$\Delta\boldsymbol{\tau} = \mathbf{g(q)} + \mathbf{J^T(q)h} - \boldsymbol{\tau} \qquad (3)$$

where $\boldsymbol{\tau}$ can be detected by F/T sensor mounted on the manipulator's end actuator.

4.3 Calculation of Gravity's Influence

Normally for a 6-DoF robotic manipulator, the influence on the robot's kinematic characteristic caused by its last 3 joints is small and can be equivalent to a single link, so the manipulator is simplified to a 3-link model, with joint names waist, shoulder and elbow, respectively. Using Lagrange's equation, $\mathbf{g(q)}$ can be calculated[10].

Fig.4 is the schema of the simplified model, in which L_1, L_2 and L_3 note for link length; $\mathbf{o_0}$, $\mathbf{o_1}$, $\mathbf{o_2}$, $\mathbf{o_3}$ note for the origins of the relative local coordinate system; m_{c1}, m_{c2}, m_{c3} note for the mass of the 3 links, L_{c1}, L_{c2}, L_{c3} note for distance between the CoMs of 3 links and its local coordinate system's origins; $\mathbf{T_1^0}$, $\mathbf{T_2^1}$, $\mathbf{T_3^2}$ denote the relative homogeneous transformation matrices.

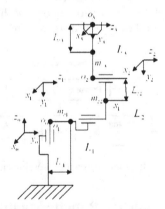

Fig. 4. Simplified Model of the Assistive Robotic Manipulator

Let h_1, h_2 and h_3 denote the height of all 3 links' CoMs in base coordinated system, then we have:

$$h_1 = [0\ 0\ 1\ 0]\mathbf{T_1^0}[0\ 0\ L_{c1}\ 1]^T$$

$$h_2 = [0\ 0\ 1\ 0]\mathbf{T_1^0 T_2^1}[0\ L_{c1}\ 0\ 1]^T$$

$$h_3 = [0\ 0\ 1\ 0]\mathbf{T_1^0 T_2^1 T_3^2}[0\ L_{c3}\ 0\ 1]^T$$

The total potential energy of the manipulator:

$$U = m_{c1}gh_1 + m_{c2}gh_2 + m_{c3}gh_3$$

So the torque generated by gravity ($\mathbf{g(q)} = [\tau_{g1} \quad \tau_{g2} \quad \tau_{g3}]^T$) are calculated by:

$$\tau_{gi} = \partial U / \partial q_i \quad i = 1,2,3$$

where q_i notes for joint position.

5 Experiments and Results Analysis

5.1 Experiment Prototype

Experiment platform is a 6-DoF assistive robotic manipulator showed in Fig.5. Equivalent 3 link model's link length and mass information is listed in Tab.1. Torque sensors are mounted on waist, shoulder and elbow.

Table 1. Link parameters

Link	L1	L2	L3
Length (mm)	0.208	0.450	0.550
Mass (kg)	1.55	1.80	2.00

5.2 Result Analysis

The control cycle for collision detection is 18 ms, and the manipulator is able to respose to the collision in the next cycle. The torque threshold in this experiement is set to 1Nm.

The manipulator is set to have a collision with a balloon and two sets of the experiment were carried out under rotation speed of 10 ° /sec and 30 ° /sec so as to give a comparison. Fig.5 shows the snapshots of the experiment.

Fig. 5. Snapshots of collision experiment

Fig.6 illustrates the speed variation $\dot{q}$ and torque diviation $\Delta\tau_3$ of elbow when a collision happens. The curves fluctuate a little when the manipulator works properly, the faint fluctuation can be caused by the following two reasons:

1. Errors generated by the torque sensors, mainly because of the temperature drifting and time-lag dynamic error.

2. Errors caused by simplifying the dynamic model of the assistive manipulator. Influence of the velocity, accecelaration, friction is neglected, thus the faster it operates, the bigger the error generates. Besides, the electric noise caused by wiring could also be a subtle influence.

By setting a proper torque threshold, negtive influence caused by error can be eliminated, like in this experiment, 1 Nm in far from enough.

From Fig.6 one can tell that when a collision happens, the output torque of the elbow will rise and exceed the threshold, the controller will stop the motor in the next control cycle once the a collision is determined. Due to the impact on the sensors, the output signal will fluctuate a little before it returns to a normal value.

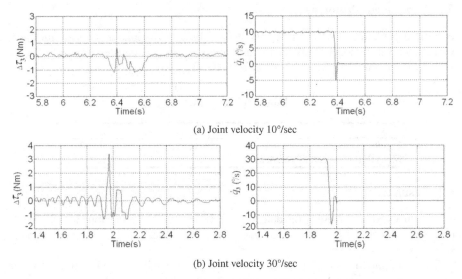

(a) Joint velocity 10°/sec

(b) Joint velocity 30°/sec

Fig. 6. Collision response of elbow with different speed

6   Conclusion

A new collision detection method for assistive robotic manipulator based on joint torque sensors is introduced in this paper. A comaprison between the actual output torque and the reference output torque is carried out to find the diviations which indicates whether a collision happens or not. In the sensor design phase, finite element analysis is carried out to find the best place to affix the strain foils to the torque disk and a signal amplification and filtering circuit with low temperature drift and high capacity of disturbance resistance is developed. Reference output torque is calculated from the manipulator's dynamic model, which is simplified to a 3-link system considering its low operational speed. Three torque sensors are distributed at three main joints of the manipulator so that the collision dection area covers the whole robot. An experiment is brought out and the result shows the effectiveness of this deisgn.

Acknowledgement. This work is partly supported by the National High Technology Research and Development Program of China under grant 2006AA040203, the Natural Science Foundation of China under grant 60775062 and 60934006, the State Key Laboratory of Robotics and System (HIT) and Research Councils UK under UK-China Science Bridge Grant No. EP/G042594/1.

References

1. Albu-Schäffer, A., Haddadin, S., Ott, C., Stemmer, A., Wimbock, T., Hirzinger, G.: The DLR lightweight robot: design and control concepts for robots in human environments. Industrial Robot Journal 34(5), 376–385 (2007)
2. Jeong, S.H., Takahashi, T., Nakano, E.: A safety service manipulator system: the reduction of harmful force by a controllable torque limiter. In: Proceedings of the IEEE/RSJ International Conference on Intelligent Robots and Systems, Sendai, Japan, pp. 162–167 (2004)
3. Kosuge, K., Matsumoto, T.: Collision detection of manipulator based on adaptive control law. In: Proceedings of the IEEE/ASME International Conference on Advanced Intelligent Mechatronics, pp. 117–122 (2001)
4. Hirzinger, G., Albu-Schäffer, A., Hähnle, M., Schaefer, I., Sporer, N.: On a new generation of torque controlled light-weight robots. In: Proceedings of the IEEE International Conference of Robotics and Automation, Seoul, Korea, pp. 3356–3363 (2001)
5. Luca, A.D., Mattone, R.: Sensorless robot collision detection and hybrid force/motion control. In: Proceedings of the IEEE International Conference on Robotics and Automation, Barcelona, Spain, pp. 999–1004 (2005)
6. Ni, F., Gao, X., Jin, M., Zhu, Y., Xu, Z.: Torque Sensor Design for Light Weight Arm Joint. Chinese Journal of Scientific Instrument 27(2), 191–195 (2006)
7. Hirzinger, G., Butterfaß, J., Fischer, M., Grebenstein, M., Hähnle, M., Liu, H., Schaefer, I., Sporer, N.: A mechatronic approach to the design of light-weight arms and multi-fingered hands. In: Proceedings of the IEEE International Conference on Robotics and Automation, San Francisco, USA, pp. 46–54 (2000)
8. Shi, S., Xie, Z., Ni, F., Liu, H.: Development of High Integration Modular Joint for Space Manipulator. Journal of Xi'an Jiaotong University 41(2), 162–166 (2007)
9. Sciaviocco, L., Siciliano, B.: Modeling and control of robot manipulators. Springer, London (2000)
10. Heinzmann, J., Zelinsky, A.: A safe-control paradigm for human-robot interaction. Journal of Intelligent and Robotic Systems 25(4), 295–310 (1999)

Adaptive Visual Servoing with Imperfect Camera and Robot Parameters

Hesheng Wang[1,2], Maokui Jiang[1,2], Weidong Chen[1,2] and Yun-hui Liu[3]

[1] Department of Automation, Shanghai Jiao Tong University, Shanghai, 200240, China
[2] State Key Laboratory of Robotics and System (HIT), Harbin 150001, China
[3] Department of Mechanical and Automation, Chinese University of Hong Kong,
Hong Kong
{wanghesheng,wdchen}@sjtu.edu.cn

Abstract. This paper presents a new adaptive controller for dynamic image-based visual servoing of a robot manipulator when the camera intrinsic and extrinsic parameters and robot physical are not calibrated. To cope with nonlinear dependence of the image Jacobian on the unknown parameters, this controller employs depth-independent image Jacobian which does not depend on the scale factors determined by the depths of feature points. By removing the scale factors, the camera and robot parameters appear linearly in the close-loop dynamics so that a new algorithm is developed to estimate these parameters on-line. Lyapunov theory is employed to prove asymptotic convergence of the image errors based on the robot dynamics. Simulations have been conducted to demonstrate the performance of the proposed controller.

Keywords: Visual servoing, Adaptive, Imperfect parameters.

1 Introduction

Visual servoing is an approach of controlling the motion of robot by visual feedback. An image-based controller usually employs control algorithm depends on the intrinsic and extrinsic parameters of the camera and robot physical parameters, calibration accuracy of those parameters significantly affect the control errors. However, calibration is tedious and costly. It is desirable to use uncalibrated visual signals directly in controller design. To this objective, various methods have been developed for estimation of the image Jacobian matrix. Hosoda and Asada [2] employed the Broyden updating formula for the estimation. Hu [3] proposed a homography-based visual servo control algorithm with imperfect camera calibration. Fang [4] also employed homography-based approach for visual regulation of mobile robots. Gans [5] developed hybrid switched-system control method for visual servoing. However, the methods mentioned above are based on kinematics only and the nonlinear forces in robot dynamics are neglected. Neglecting the nonlinear forces affect not only control errors but also the stability. The controllers proposed in [6] can cope with the unknown parameters but apply to planar manipulators only. Cheah [7] proposed a visually servoed adaptive controller for motion and force tracking with uncertainties in the constraint surface, kinematics, dynamics, and camera model. In our early work [8][9], we

K. Li et al. (Eds.): LSMS / ICSEE 2010, Part I, CCIS 97, pp. 124–131, 2010.
© Springer-Verlag Berlin Heidelberg 2010

proposed an adaptive controller for dynamic and uncalibrated visual servoing in eye-to-hand configuration. In [10][11], we extend our earlier results to an eye-in-hand configuration.

This paper presents a new controller to regulate a set of feature points on the image plane to desired positions by controlling motion of a robot manipulator. This is an extension of our work [8] by assuming both the camera and robot parameters are not known. This controller takes into account the nonlinear forces in the robot dynamics. We employ a depth-independent image Jacobian, which differs from the image Jacobian by the scale factor, to map the image errors onto the joint space of the robot. Adaptive algorithm has been developed for on-line estimation of the unknown parameters. Lyapunov theory is employed to prove the asymptotic convergence of the image errors. To verify the performance, simulations have been conducted on a 3 degrees of freedom (DOF) robot manipulator.

2 Kinematics and Dynamics

In this work, we consider a fixed camera set-up (Fig.1). Assume that the robot physical parameters, the feature points positions as well as camera intrinsic parameters and the extrinsic parameters are not calibrated. The problem is defined as:

Problem 1. Design a proper joint input for the robot such that the feature points asymptotically track the desired trajectories in uncalibrated environments.

Denote the joint angle of the manipulator by a $n \times 1$ vector $\mathbf{q}(t)$, where n is the number of the joints of the manipulator. Denote the *perspective projection matrix* of the camera by $\mathbf{M} \in \mathfrak{R}^{3 \times 4}$. Denote the position of the tracked point in the end-effector frame as $\mathbf{r}$, which is unknown. Denote the image coordinates of feature point on the image plane by $\mathbf{y}(t) = (u_i, v_i)^T$ and its coordinates with respect to the robot base frame by a 3×1 vector $\mathbf{x}(t)$, then

$$\mathbf{x}(t) = \mathbf{R}(\mathbf{q}(t))\mathbf{r} + \mathbf{p}(\mathbf{q}(t)) \tag{1}$$

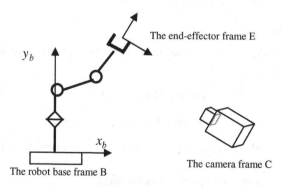

Fig. 1. A fixed camera setup for visual tracking

where $\mathbf{R}(\mathbf{q}(t))$ is the 3 by 3 rotation matrix and $\mathbf{p}(\mathbf{q}(t))$ denotes the 3 by 1 translational vector from the end-effector frame to the robot base frame. Let $\tilde{\mathbf{J}}(\mathbf{q}(t)) \in \mathfrak{R}^{6 \times n}$ denote the manipulator Jacobian matrix. From the forward kinematics of the manipulator, we obtain

$$\dot{\mathbf{x}}(t) = \left(\mathbf{I}_{3 \times 3} \quad -sk(\mathbf{R}(\mathbf{q}(t))\mathbf{r})\right)\tilde{\mathbf{J}}(\mathbf{q}(t))\dot{\mathbf{q}}(t) \tag{2}$$

where sk is a skew-symmetric matrix operator. Under the perspective projection model, the coordinates are related by

$$\begin{pmatrix} \mathbf{y}(t) \\ 1 \end{pmatrix} = \frac{1}{{}^c z(t)} \mathbf{M} \begin{pmatrix} \mathbf{x}(t) \\ 1 \end{pmatrix} \tag{3}$$

where ${}^c z(t)$ denotes the depth of the feature point with respect to the camera frame:

$$^c z(\mathbf{q}(t)) = \mathbf{m}_3^T \mathbf{x}(t) + m_{34} \tag{4}$$

where $\mathbf{m}_i^T$ denotes the first three components of the i-th row vector of the perspective projection matrix $\mathbf{M}$. Differentiating (4) results in

$$^c \dot{z}(\mathbf{q}(t)) = \underbrace{\mathbf{m}_3^T \left(\mathbf{I}_{3 \times 3} \quad -sk(\mathbf{R}(\mathbf{q}(t))\mathbf{r}_i)\right)\tilde{\mathbf{J}}(\mathbf{q}(t))}_{\mathbf{d}(t)}\dot{\mathbf{q}}(t) = \mathbf{d}(t)\dot{\mathbf{q}}(t) \tag{5}$$

By differentiating (3), we obtain the following relation:

$$\dot{\mathbf{y}}(t) = \frac{1}{{}^c z(t)} \mathbf{D}(t)\dot{\mathbf{q}}(t) \tag{6}$$

where $\mathbf{D}(t)$ is the *depth-independent image Jacobian matrix*, i.e.:

$$\mathbf{D}(t) = \begin{pmatrix} \mathbf{m}_1^T - u_i(t)\mathbf{m}_3^T \\ \mathbf{m}_2^T - v_i(t)\mathbf{m}_3^T \end{pmatrix}\left(\mathbf{I}_{3 \times 3} \quad -sk(\mathbf{R}(\mathbf{q}(t))\mathbf{r}_i)\right)\tilde{\mathbf{J}}(\mathbf{q}(t)) \tag{8}$$

It is important to note :

Property 2. *For any* 4×1 *vector* $\boldsymbol{\rho}$, *the product* $\mathbf{D}(t)\boldsymbol{\rho}$ *and* $\mathbf{d}(t)\boldsymbol{\rho}$ *can be written in the following linear form:*

$$\mathbf{D}(t)\boldsymbol{\rho} = \mathbf{B}(\boldsymbol{\rho}, \mathbf{y}_i(t))\boldsymbol{\theta}$$

$$\mathbf{d}(t)\boldsymbol{\rho} = \mathbf{b}(\boldsymbol{\rho}, \mathbf{y}_i(t))\boldsymbol{\theta} \tag{9}$$

where $\mathbf{B}(\boldsymbol{\rho}, \mathbf{y}_i(t))$ *and* $\mathbf{b}(\boldsymbol{\rho}, \mathbf{y}_i(t))$ *are regression matrices. Vector* $\boldsymbol{\theta}$ *includes all the combined unknown parameters.*

The dynamic equation of a manipulator has the form:

$$\mathbf{H}(\mathbf{q}(t))\ddot{\mathbf{q}}(t) + [\frac{1}{2}\dot{\mathbf{H}}(\mathbf{q}(t)) + \mathbf{C}(\mathbf{q}(t), \dot{\mathbf{q}}(t))]\dot{\mathbf{q}} + \mathbf{g}(\mathbf{q}(t)) = \boldsymbol{\tau} \tag{10}$$

where $\mathbf{H}(\mathbf{q}(t))$ is the inertia matrix, $\mathbf{C}(\mathbf{q}(t),\dot{\mathbf{q}}(t))$ is a skew-symmetric matrix The term $\mathbf{g}(\mathbf{q}(t))$ represents the gravitational force, and τ is the joint input of the robot manipulator.

3 Adaptive Visual Servoing

3.1 Controller Design

Denote the desired position of the feature point on the image plane by $\mathbf{y}_d$, which is a constant vector. The image error is obtained by measuring the difference between the current position and the desired one:

$$\Delta \mathbf{y}(t) = \mathbf{y}(t) - \mathbf{y}_d \tag{11}$$

Denote the time-varying estimation of the unknown parameters θ by $\hat{\theta}(t)$. Using the estimated parameters, we propose the following controller:

$$\tau = \hat{\mathbf{g}}(\mathbf{q}(t)) - \mathbf{K}_1\dot{\mathbf{q}}(t) - (\hat{\mathbf{D}}^T(\mathbf{y}(t)) + \frac{1}{2}\hat{\mathbf{d}}^T(t)\Delta \mathbf{y}^T(t))\mathbf{B}\Delta \mathbf{y}(t) \tag{12}$$

The matrix $\hat{\mathbf{D}}(\mathbf{y}(t))$ represents an estimation of the depth-independent interaction matrix calculated using the estimated parameters. $\mathbf{K}_1$ is a $n \times n$ positive-definite velocity gain matrix and $\mathbf{B}$ is positive definite position gain matrix.

By substituting the control law (12) into the robot dynamics (10), we obtain the following closed loop dynamics:

$$\mathbf{H}(\mathbf{q}(t))\ddot{\mathbf{q}}(t) + (\frac{1}{2}\dot{\mathbf{H}}(\mathbf{q}(t)) + \mathbf{C}(\mathbf{q}(t),\dot{\mathbf{q}}(t)))\dot{\mathbf{q}}(t) =$$

$$(\hat{\mathbf{g}}(\mathbf{q}(t)) - \mathbf{g}(\mathbf{q}(t))) - \mathbf{K}_1\dot{\mathbf{q}}(t) - (\mathbf{D}^T(\mathbf{y}(t)) + \frac{1}{2}\mathbf{d}^T(t)\Delta \mathbf{y}^T(t))\mathbf{B}\Delta \mathbf{y}(t) \tag{13}$$

$$-\left((\hat{\mathbf{D}}^T(\mathbf{y}(t)) - \mathbf{D}^T(\mathbf{y}(t))) + \frac{1}{2}(\hat{\mathbf{d}}^T(t) - \mathbf{d}^T(t))\Delta \mathbf{y}^T(t)\right)\mathbf{B}\Delta \mathbf{y}(t)$$

From the property of robot dynamics, the first term can be represented as a linear form of the estimation errors of the robot parameters as follows:

$$(\hat{\mathbf{g}}(\mathbf{q}(t)) - \mathbf{g}(\mathbf{q}(t))) = \mathbf{U}(\mathbf{q}(t),\mathbf{y}(t))\Delta \theta_g(t) \tag{14}$$

where $\Delta \theta_g(t) = \hat{\theta}_g(t) - \theta_g$, representing the estimation error and $\mathbf{U}(\mathbf{q}(t),\mathbf{y}(t))$ is a regressor matrix without depending on the unknown parameters.

Similar, from the Property 1, we can obtain:

$$-\left((\hat{\mathbf{D}}^T(\mathbf{y}(t)) - \mathbf{D}^T(\mathbf{y}(t))) + \frac{1}{2}(\hat{\mathbf{d}}^T(t) - \mathbf{d}^T(t))\Delta \mathbf{y}^T(t)\right)\mathbf{B}\Delta \mathbf{y}(t) = \mathbf{Y}(\mathbf{q}(t),\mathbf{y}(t))\Delta \theta(t) \tag{15}$$

where $\Delta\theta(t) = \hat{\theta}(t) - \theta$ representing the estimation error and $Y(q(t), y(t))$ is a regressor matrix without depending on the unknown parameters.

3.2 Parameter Estimation

In this subsection, we propose an adaptive algorithm to estimate the unknown parameters on-line. The basic idea is to combine the Slotine-Li algorithm with an on-line minimization of the estimated projection error of the feature point. From (3), the estimated projection error of the feature point at this time instant is given by

$$e(t_j, t) = y(t_j)^c\hat{z}^T(t) - \hat{P}(t)\begin{pmatrix} x(\hat{r}, q(t_j)) \\ 1 \end{pmatrix} = W(x(t_j), y(t_j))\Delta\theta(t) \qquad (16)$$

It should be noted that the error $e(t_j, t)$ can be calculated from the measurement of the 3-D position and projection of the feature point without knowing the true parameters. On the trajectory of the feature point, we can select m such positions, and hence m equations like (16) can be obtained. In the parameters adaptation, we update the estimation in the direction of reducing the estimated projection errors $e(t_j, t)$. Following is the adaptive rule for updating the estimation of the parameters:

$$\frac{d}{dt}\hat{\theta}(t) = -\Gamma_1^{-1}\{Y^T(q(t), y(t))\dot{q}(t) + \sum_{j=1}^m W^T(x(t_j), y(t_j))K_3 e(t_j, t)\} \qquad (17)$$

$$\frac{d}{dt}\hat{\theta}_g(t) = -\Gamma_2^{-1}U^T(q(t), y(t))\dot{q}(t) \qquad (18)$$

where Γ_1, Γ_2 and K_3 are positive-definite and diagonal gain matrices.

3.3 Stability Analysis

Theorem 1. *Under the control of the proposed controller (12) and the adaptive algorithm (17)(18) for parameters estimation, the image error of the feature point is convergent to zero, i.e.*

$$\lim_{t\to\infty} \Delta y(t) = 0 \qquad (19)$$

Proof. Introduce the following non-negative function:

$$v(t) = \frac{1}{2}\{\dot{q}^T(t)H(q(t))\dot{q}(t) + {}^c z\Delta y^T(t)B\Delta y(t) + \Delta\theta^T(t)\Gamma_1\Delta\theta(t) + \Delta\theta_g^T(t)\Gamma_2\Delta\theta_g(t)\} \quad (20)$$

Multiplying the $\dot{q}^T(t)$ from the left to the closed loop dynamics (13) results in

$$\dot{q}^T(t)H(q(t))\ddot{q}(t) + \frac{1}{2}\dot{q}^T(t)\dot{H}(q(t))\dot{q}(t) = -\dot{q}^T(D^T(y(t)) + \frac{1}{2}d^T(t)\Delta y^T(t))B\Delta y(t)$$
$$-\dot{q}^T(t)K_1\dot{q}(t) + \dot{q}^T(t)Y(q(t), y(t))\Delta\theta(t) + \dot{q}^T(t)U(q(t), y(t))\Delta\theta(t) \qquad (21)$$

From equations (6) and (8), we have

$$\dot{\mathbf{q}}^T(t)\mathbf{D}^T(\mathbf{y}(t)) = {}^c z(\mathbf{q}(t))\dot{\mathbf{y}}^T(t) = {}^c z(\mathbf{q}(t))\Delta\dot{\mathbf{y}}^T(t) \tag{22}$$

By multiplying the $\Delta\boldsymbol{\theta}^T(t)$ from the left to the adaptive rule (17)(18), we obtain

$$\Delta\boldsymbol{\theta}^T(t)\boldsymbol{\Gamma}_2\Delta\dot{\boldsymbol{\theta}}(t) = -\Delta\boldsymbol{\theta}(t)\mathbf{Y}^T(\mathbf{q}(t),\mathbf{y}(t))\dot{\mathbf{q}}(t) - \sum_{j=1}^{m}\Delta\boldsymbol{\theta}^T(t)\mathbf{W}^T(t_j)\mathbf{K}_3\mathbf{W}(t_j)\Delta\boldsymbol{\theta}(t) \tag{24}$$

$$\Delta\boldsymbol{\theta}_g^T(t)\boldsymbol{\Gamma}_2\Delta\dot{\boldsymbol{\theta}}_g(t) = -\Delta\boldsymbol{\theta}_g(t)\mathbf{U}^T(\mathbf{q}(t),\mathbf{y}(t))\dot{\mathbf{q}}(t) \tag{25}$$

Differentiating the function $v(t)$ in (20) results in

$$\dot{v}(t) = \dot{\mathbf{q}}^T(t)(\mathbf{H}(\mathbf{q}(t))\ddot{\mathbf{q}}(t) + \frac{1}{2}\dot{\mathbf{H}}(\mathbf{q}(t))\dot{\mathbf{q}}(t)) + {}^c z(\mathbf{q}(t))\Delta\mathbf{y}^T(t)\mathbf{B}\Delta\dot{\mathbf{y}}(t)$$

$$+\frac{1}{2}{}^c\dot{z}(\mathbf{q}(t))\Delta\mathbf{y}^T(t)\mathbf{B}\Delta\mathbf{y}(t) + \Delta\boldsymbol{\theta}^T(t)\boldsymbol{\Gamma}_1\Delta\dot{\boldsymbol{\theta}}(t) + \Delta\boldsymbol{\theta}_g^T(t)\boldsymbol{\Gamma}_2\Delta\dot{\boldsymbol{\theta}}_g(t) \tag{26}$$

By combining the equations (22)-(26), we have

$$\dot{v}(t) = -\dot{\mathbf{q}}^T(t)\mathbf{K}_1\dot{\mathbf{q}}(t) - \sum_{j=1}^{m}\Delta\boldsymbol{\theta}^T(t)\mathbf{W}^T(t_j)\mathbf{K}_3\mathbf{W}(t_j)\Delta\boldsymbol{\theta}(t) \tag{27}$$

From the Barbalat's Lemma, we conclude that

$$\lim_{t\to\infty}\dot{\mathbf{q}}(t) = \mathbf{0}$$
$$\lim_{t\to\infty}\mathbf{W}(\mathbf{x}(t_j),\mathbf{y}(t_j))\Delta\boldsymbol{\theta} = \mathbf{0} \tag{28}$$

By considering the invariant set of the system when $\dot{V}(t) = 0$, we can conclude the convergence of the position error of the target projections on the image plane to zero

4 Simulations

In this section, we show the performance of the proposed image-based controller by simulations. We conducted the simulations on a 3 DOF manipulator. The physical parameters of the robot are set as $l_1 = 0.24m$, $l_2 = 0.43m$, $l_3 = 0.09m$. $m_1 = 17.4kg$, $m_2 = 4.8kg$. The control gains are $\mathbf{K}_1 = 4$, $\mathbf{B} = 0.00005$, $\boldsymbol{\Gamma}_1 = 20000$ $\boldsymbol{\Gamma}_2 = 5000$. The real camera intrinsic parameters are $a_u = 800$, $a_v = 900$, $u_0 = 300$, $v_0 = 400$. The initial estimated values are $\hat{a}_u = 900$, $\hat{a}_v = 920$, $\hat{u}_0 = 350$, $\hat{v}_0 = 490$. The initial estimated homogenous transformation matrix of the camera frame with respect to the end-effector frame is $\hat{\mathbf{T}}(0) = \begin{bmatrix} 0 & 1 & 0 & 0.003 \\ 0.3090 & 0 & 0.9511 & -0.005 \\ 0.9511 & 0 & -0.3090 & 0.007 \\ 0 & 0 & 0 & 1 \end{bmatrix}$.

As shown in Figure 2(a), the image feature points asymptotically converge to the desired ones. Figure 2(b) plot the trajectory of the feature on the image plane. The results confirmed the convergence of the image error to zero under control of the proposed method.

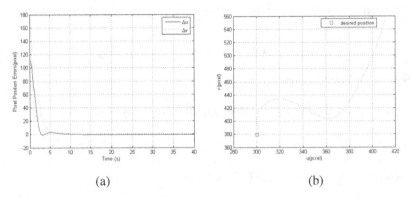

(a)   (b)

Fig. 2. (a) Position errors of the image feature; (b) Trajectory on the image plane

5   Conclusions

In this paper, we proposed a new controller to regulate a set of feature points on the image plane to desired positions by controlling motion of a robot manipulator when both the camera and robot parameters are not known. The controller employs a depth-independent image Jacobian, which differs from the image Jacobian by the scale factor, to map the image errors onto the joint space of the robot. Adaptive algorithm has been developed for on-line estimation of the unknown parameters. Lyapunov theory is employed to prove the asymptotic convergence of the image errors based on the robot dynamics. Simulations have been performed to verify the performance of the proposed method.

Acknowledgments. This work is partly supported by the National High Technology Research and Development Program of China under grant 2006AA040203, the Natural Science Foundation of China under grant 60775062 and 60934006, the State Key Laboratory of Robotics and System (HIT) and Research Councils UK under UK-China Science Bridge Grant No. EP/G042594/1.

References

1. Hutchinson, S., Hager, G.D., Corke, P.I.: A tutorial on visual servo control. IEEE Tran. Robotics and Automation 12, 651–670 (1996)
2. Hosada, K., Asada, M.: Versatile Visual Servoing without knowledge of True Jacobain. In: IEEE/RSJ Int. Conf. on Intelligent Robots and Systems, pp. 186–191 (1994)
3. Hu, G., MacKunis, W., Gans, N., Dixon, W.E., Chen, J., Behal, A., Dawson, D.: Homography-Based Visual Servo Control with Imperfect Camera Calibration. IEEE Trans. Automatic Control 54, 1318–1324 (2009)

4. Fang, Y., Dixon, W., Dawson, D., Chawda, P.: Homography-based visual servo regulation of mobile robots. IEEE Trans. Syst., Man Cybern. 35, 1041–1050 (2005)
5. Gans, N.R., Hutchinson, S.A.: Stable Visual Servoing Through Hybrid Switched-System Control. IEEE Trans. Robotics 23, 530–540 (2007)
6. Kelly, R., Carelli, R., Nasisi, O., Kuchen, B., Reyes, F.: Stable Visual Servoing of Camera-in-Hand Robotic Systems. IEEE/ASME Trans. Mechatronics 5, 39–48 (2000)
7. Cheah, C.C., Hou, S.P., Zhao, Y., Slotine, J.-J.E.: Adaptive Vision and Force Tracking Control for Robots With Constraint Uncertainty. IEEE/ASME Trans. Mechatronics 15, 389–399 (2010)
8. Liu, Y.H., Wang, H., Wang, C., Lam, K.: Uncalibrated Visual Servoing of Robots Using a Depth-Independent Image Jacobian Matrix. IEEE Trans. Robotics 22, 804–817 (2006)
9. Wang, H., Liu, Y.H., Zhou, D.: Dynamic Visual Tracking for Manipulators Using An Uncalibrated Fixed Camera. IEEE Trans. Robotics 23, 610–617 (2007)
10. Wang, H., Liu, Y.H., Zhou, D.: Adaptive Visual Servoing Using Point and Line Features with An Uncalibrated Eye-in-hand Camera. IEEE Tran. Robotics 24, 843–857 (2008)
11. Wang, H., Liu, Y.H.: A New Approach to Dynamic Eye-in-hand Visual Tracking Using Nonlinear Observers. IEEE/ASME Trans. Mechatronics 15 (2010)

An Interactive Method to Solve the Priorities of Attributes While the Preferences of Evaluated Units Are under Considering*

Guohua Wang[1], Jingxian Chen[1,**], Qiang Guo[2], and Liang Liang[3]

[1] School of Business, Nantong University, Nantong, 226019, P.R. China
Tel.: +86-13962967110; Fax: +86-513-8501 2561
jxchenms@163.com
[2] Tourism College, Hainan University, Haikou, 570228, P.R. China
[3] School of Management, University of Science and Technology of China,
Hefei, 230026, P.R. China

Abstract. Formerly, the preferences or attitudes of evaluated units are not under considering while the group decision methods are applied in evaluations. In this paper, a new model is proposed for solving the priorities of attributes. By simulating the decision-making process, the model uses an interactive method to optimize the priorities of attributes and determine the scores of evaluated units. An applied example given at the end of the paper shows the process of this method.

Keywords: Group decision; Decision analysis; Linear programming; Subjective evaluation; Objective evaluation.

1 Introduction

In recent years, people have been paying more attention to the study of group decision, which is a major study in the field of decision science. There are generally two kinds of system evaluation methods, subjective evaluation, objective evaluation. For example, Delphi method and Analytic Hierarchy Process belong to subjective evaluation methods, Essential Component Analysis, Data Envelopment Analysis and Hierarchical Clustering Methods belong to the latter.[1-5] But it is a matter of the methods mentioned hereinbefore that the preferences or attitudes of evaluated units are not under considering. In a practical project, the evaluation of the ability of technical innovation in 10 provinces of China, we found that every evaluated unit (decision-making unit, DMU) wanted to show his attitude and put it into the process of the evaluation. Because of the differences in many aspects, we must take the preferences of DMUs into consideration. Therefore in this practical project we set up a new model to enable the units to give their opinions.

* This work was partially by the National Natural Science Foundation of China (NSFC) under grant No. 70761001 and by the Nantong University Talents Introduction Foundation under grant No. 08R05 and by the Nantong University Social Science Foundation under grant No. 09W021.
** Corresponding author.

K. Li et al. (Eds.): LSMS / ICSEE 2010, Part I, CCIS 97, pp. 132–143, 2010.

This model has a system of indices for the evaluation. The system can be determined either by the director or by both the director and the DMUs. You can also make decisions by regular rules [6,7]. The director will determine some restrictions and rules, which can be used by the DMUs to solve their own optimal priorities of attributes (making up of priority vectors) according to their preferences and the balance among DMUs [8]. Then the priority vectors are handed to the director, which would synthesize all these priority vectors and obtain the synthesized reference priority vector. When the reference priority vector is fed back to each DMU, the unit will solve his own priority vector once more and hand over the new optimal priority vectors to the director. Continue this kind of feedback until we meet our demands. Finally the director solves the optimal synthesized priority vector, which can be used to determine the evaluation of each DMU [9,10].

In this kind of decision-making model, every DMU would give the priority vector by himself based on the restrictions given by the director. For every DMU tends to emphasize his own superiority he would give bigger priorities to the attributes of his advantages. So the priority vector of every DMU is different from others and so does the evaluating results. Therefore, every unit should make the optimal priority vector under the influence of other units in the evaluating system.

The process is listed as the following figure 1.

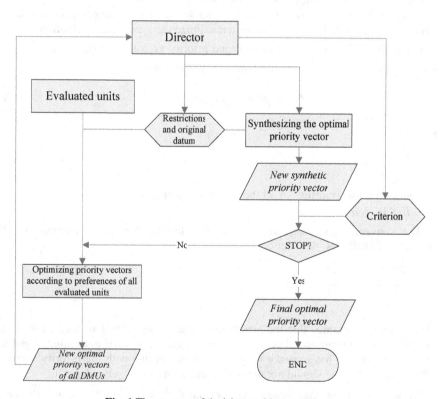

Fig. 1 The process of decision-making model

2 Optimization Model of Evaluated Units

Suppose n DMUs with m attributes are concerned with the evaluation. Let

$$A = [x_{ij}]_{n \times m} = \begin{bmatrix} x_{11} & x_{12} & \cdots & x_{1m} \\ x_{21} & x_{22} & \cdots & x_{2m} \\ \vdots & & \ddots & \vdots \\ x_{n1} & x_{n2} & \cdots & x_{nm} \end{bmatrix}$$

where x_{ij} (suppose it has been normalized) denotes the jth attribute value of the ith unit.

Let w_{ij} be the weight of jth attributes in the ith unit.

Let lb and ub be the lower and upper bound vectors.

Let $W^{(k)} = \left(w_1^{(k)}, w_2^{(k)}, \cdots, w_m^{(k)} \right)^T$ be the synthesized priority vector ($W^{(1)}$ is the original priority vector) in the kth cycle (See section 3, 4, 5).

Every DMU has his own preference and tends to give bigger priorities to his advantages. In another word, he tends to maximize the synthetical evaluation.

For each DMU, we can get the optimal priority vector, $W_i = \left(w_{i1}, w_{i2}, \cdots, w_{im} \right)^T$, by the following optimization problem (utility function):

$$Z_i = \max \sum_{j=1}^{m} w_{ij} x_{ij} \tag{1}$$

Subject to

a. $\sum_{j=1}^{m} w_{ij} = 1$

(The sum of the priorities of all the attributes must be 1.)

b. $lb \leq W_i \leq ub$

(For convenience's sake, we suppose the director has set restrictions for the priorities of attributes. See section 4)

c. $\sum_{j=1}^{m} w_{ij} x_{ij} \geq \sum_{j=1}^{m} w_j^{(k)} x_{ij}$

for all $i = 1, 2, \cdots, n$.

(This restriction ensures that the evaluation values of all DMUs won't be lower than that from the last synthesized priority vector. When every DMU goes after its greatest evaluation value, it has to make its priority vector in line with the priority vector of other DMUs. Anyone should not put undue stress on its own advantages.)

So, we have n optimum priority vectors, i.e. $W_i^{(k)} = \left(w_{i1}^{(k)}, w_{i2}^{(k)}, \cdots, w_{im}^{(k)} \right)^T$, for all $i = 1, 2, \cdots, n$.

Then, we can synthesize the n optimal priority vectors of all units, and obtain the revised priority vector (See section 3):

$$W^{(k+1)} = \left(w_1^{(k+1)}, w_2^{(k+1)}, \cdots, w_m^{(k+1)}\right)^T$$

3 Method of Synthesizing Priority Vectors

Here is the method: construct a priority vector denoted by W^*, which is the closest to the priority vectors of the units as a whole. The included angles between the vectors indicate how close the vectors are.

Let $W_i = \left(w_1^i, w_2^i, \cdots, w_m^i\right)^T$, $W_j = \left(w_1^j, w_2^j, \cdots, w_m^j\right)^T$. The included angle between W_i and vector W_j is:

$$\theta_{i,j} = \arccos \frac{(W_i, W_j)}{|W_i||W_j|} = \arccos \frac{\sum_{k=1}^{n} w_k^i w_k^i}{\sqrt{\sum_{k=1}^{n} \left(w_k^i\right)^2 \sum_{k=1}^{n} \left(w_k^j\right)^2}} \tag{2}$$

where $(W_i, W_j) = |W_i||W_j| \cos \theta_{ij}$ represents the product of vectors, i.e.

$$\cos \theta_{i,j} = \frac{(W_i, W_j)}{|W_i||W_j|} = \frac{\sum_{k=1}^{n} w_k^i w_k^i}{\sqrt{\sum_{k=1}^{n} \left(w_k^i\right)^2 \sum_{k=1}^{n} \left(w_k^j\right)^2}} \tag{3}$$

Let $W^* = \left(w_1^*, w_2^*, \cdots, w_m^*\right)^T$. And the lower and upper bound vectors can be denoted by lb and ub.

Then we can obtain the synthesized priority vector W* by the following optimization problem:

$$Z^* = \max \sum_{i=1}^{n} \cos \theta_i^* = \max \sum_{i=1}^{n} \frac{(W^*, W_i)}{|W^*||W_i|} \tag{4}$$

Subject to

$$\sum_{j=1}^{m} w_i^* = 1$$

$$lb \leq W^* \leq ub$$

4 The Bound Vectors and the Original Priority Vector $W^{(1)}$

We can assume that the director invites l experts to participate in the estimation of the priorities in the beginning and obtains l priority vectors (See Table 1, suppose being normalized).

Table 1. The l priority vectors given by the experts in the first instance

Experts	Priority of attribute 1	Priority of attribute 2	$\cdots$	Priority of attribute m
1	w_{11}^0	w_{12}^0	$\cdots$	w_{1m}^0
2	w_{21}^0	w_{22}^0	$\cdots$	w_{2m}^0
$\vdots$	$\vdots$	$\vdots$	$\ddots$	$\vdots$
1	w_{l1}^0	w_{l2}^0	$\cdots$	w_{lm}^0

This leads to the lower and upper bound vectors (See Table 2).

Table 2. The lower and upper bound vectors

Bound vectors	Priority of attribute 1	Priority of attribute 2	$\cdots$	Priority of attribute m
lb	$\min\{w_{i1}^0\}$	$\min\{w_{i2}^0\}$	$\cdots$	$\min\{w_{im}^0\}$
ub	$\max\{w_{i1}^0\}$	$\max\{w_{i2}^0\}$	$\cdots$	$\max\{w_{im}^0\}$

According to formula (4), we can synthesize the 1 priority vectors given by the experts, and obtain the revised priority vector, namely the original priority vector $W(1)$. The original priority vector can be for use of the first cycle of the optimizing process.

5 Arithmetic of the Model

5.1 Discriminant of Ending the Process

Definition 5.1. Let $W^{(k+1)}$ and $W^{(k)}$ be priority vectors, the distance between $W^{(k+1)}$ and $W^{(k)}$ can be denoted by

$$d(k+1,k) = \sqrt{\sum_{j=1}^{m}\left(w_j^{(k+1)} - w_j^k\right)^2} \ .$$

Let ξ be a given constant and $\xi > 0$.

If $d(k+1,k) < \xi$, $W^{(k+1)}$ is considered the final objective priority vector. It brings to an end.

Or else, $W^{(k+1)}$ can be used as a substitute for $W^{(k)}$ in the $(k+1)^{th}$ optimizing process.

5.2 Steps of the Arithmetic

Step 1: To give a constant ξ and $\xi > 0$;
Step 2: To estimate the lower and upper bound vectors and the original priority vector $W^{(1)}$ base on the original priorities given by experts;

Step 3: To optimize the every DMU and get the optimal priorities base on the lower and upper bound vectors and the original priority vector $W^{(1)}$;

Step 4: To synthesize the optimal priorities of all DMUs base on formula (4) and get a new synthetic priority vector $W^{(2)}$,   and to judge whether the distance between $W^{(1)}$ and $W^{(2)}$ meet $d(k+1,k)<\xi$;

Step 5: If $d(k+1,k)<\xi$, $W^{(2)}$ is the final optimal priority vector and brings to an end; otherwise replace $W^{(1)}$ with $W^{(2)}$ go to step 3, loop until $d(k+1,k)<\xi$;

Step 6: Get the final optimal priority vector $W^{(k+1)}$.

5.3 The Derivation of the Model

Lemma 5.1. Given a minimum constant $\xi > 0,$ there must be an optimal solution $W^{(k+1)}$ meet $d(k+1,k)<\xi$.

Proof. For $W^{(1)}$ is a feasible solution of the model, under the constraint condition of the model the feasible regionφ^1 of the objective function must has boundary, so the objective function must has an optimal solution.

Given $W^{(1)}$, assume the optimal solution of unit i is

$$W_i^{(1)} = \left(w_{i1}^{(1)}, w_{i2}^{(1)}, \cdots, w_{im}^{(1)}\right)^T, \text{ for all } i = 1,2,\cdots,n .$$

Get the synthesized priority vector ($W^{(2)}$)) by formula (4). $W^{(1)}$ apparently is in the feasible regionφ^1.

And if $d(k+1,k)\geq\xi$,   replace $W^{(1)}$ with $W^{(2)}$, the feasible regionφ^2 of the objective function meet:

$$\varphi^2 \subseteq \varphi^1.$$

Resolve the optimal solutions of all units in the feasible regionφ^2. We can get:

$$W_i^{(2)} = \left(w_{i1}^{(2)}, w_{i2}^{(2)}, \cdots, w_{im}^{(2)}\right)^T, \text{ for all } i = 1,2,\cdots,n .$$

Get the synthesized priority vector ($W^{(3)}$) by formula (4), in the same way the feasible region:

$$\varphi^3 \subseteq \varphi^2 \subseteq \varphi^1.$$

Therefore if only $d(k,k-1)\geq\xi$, replace $W^{(k-1)}$ with $W^{(k)}$, solve the optimal solution of unit i in the feasible regionφ^k, in the same way we can get that the feasible regionφ^{k+1} meet:

$$\varphi^{k+1} \subseteq \varphi^k \subseteq \cdots\cdots \subseteq \varphi^1 \subseteq \varphi^0$$

Viz. every time the solving process repeats, the feasible region of the solution is reduced. Because the feasible region has boundary, for a given minimal constant $\xi >$ 0 there must be the optimal solution $W^{(k+1)}$ of the system. It meet: $d(k+1,k) < \xi$, k $\in$ N. That is to say, the model must be convergent.

6 An Applied Example

Now people give more and more attention on the technical innovation. As we all know, the enterprise is the subject of technical innovation and it's ability to innovate can be rated by many aspects. How to evaluate an enterprise's ability to innovate is a decision problem involving multi-attributes. In this example, we apply the method given in this paper to evaluate the ability of technical innovation of the enterprises of 10 provinces according the general principle of indicator system. Here is the table of the attributes of evaluation system:

Table 3. The attributes of evaluation system

Attributes	A/B	C/D	E/F	G/H	I/J	K	L/M	N	O	P	Q	R/S
Units	(%)	(%)	(%)	(%)	(%)	(%)	(%)					(%)
Beijing	12.0	37.0	10.8	10.3	23.5	76.00	43.65	192225.6	34704.9	237	111	62.18
Shanghai	23.7	20.7	22.4	10.2	14.2	71.68	54.46	815420.5	255020.0	336	201	70.20
Jiangsu	18.6	39.3	17.8	9.7	16.8	74.09	60.24	772167.6	244858.5	1056	369	70.51
Zhejiang	15.2	34.1	14.2	8.1	11.9	76.85	62.41	854827.7	251689.6	446	322	73.23
Anhui	9.2	37.7	8.5	8.8	14.5	61.48	50.34	238860.2	51465.0	296	63	56.71
Fujian	17.8	34.8	15.8	10.2	14.7	68.14	42.41	160478.4	24941.4	126	26	60.06
Jiangxi	8.4	30.7	8.8	7.8	6.3	69.98	49.75	72859.6	9910.3	143	11	61.16
Shandong	11.7	28.3	11.2	8.5	7.6	71.72	49.78	785531.1	243097.5	790	645	71.97
Henan	7.3	33.0	7.3	7.7	5.7	78.06	43.95	355215.6	80556.8	422	216	50.37
Guangdong	15.5	23.8	12.7	9.2	11.1	81.75	60.25	474045.2	119302.9	574	446	66.14

Table 4. The standardized original data matrix

0.50633	0.94148	0.48214	1.00000	1.00000	0.92966	0.69941	0.22487	0.13609	0.22443	0.17209	0.84911
1.00000	0.52672	1.00000	0.99029	0.60426	0.87682	0.87262	0.95390	1.00000	0.31818	0.31163	0.95862
0.78481	1.00000	0.79464	0.94175	0.71489	0.90630	0.96523	0.90330	0.96015	1.00000	0.57209	0.96286
0.64135	0.86768	0.63393	0.78641	0.50638	0.94006	1.00000	1.00000	0.98694	0.42235	0.49922	1.00000
0.38819	0.95929	0.37946	0.85437	0.61702	0.75205	0.80660	0.27942	0.20181	0.28030	0.09767	0.77441
0.75105	0.88550	0.70536	0.99029	0.62553	0.83352	0.67954	0.18773	0.09780	0.11932	0.04031	0.82016
0.35443	0.78117	0.39286	0.75728	0.26809	0.85602	0.79715	0.08523	0.03886	0.13542	0.01705	0.83518
0.49367	0.72010	0.50000	0.82524	0.32340	0.87731	0.79763	0.91894	0.95325	0.74811	1.00000	0.98279
0.30802	0.83969	0.32589	0.74757	0.24255	0.95486	0.70421	0.41554	0.31588	0.39962	0.33488	0.68783
0.65401	0.60560	0.56696	0.89320	0.47234	1.00000	0.96539	0.55455	0.46782	0.54356	0.69147	0.90318

The meanings of attributes are as follows.

A: innovative production value
B: total industrial production value
C: the number of the enterprises that have technical development divisions
D: the number of enterprises
E: new products sales income
F: total products sales income
G: technical staff
H: all employees
I: cybernation equipments
J: all equipments
K: self-financing rate
L: the expense of technical development
M: the expense to develop new products
N: the expense of technical innovation (ten thousand yuan)
O: the expense of technology import (ten thousand yuan)
P: the number of technical development divisions
Q: the number of patents
R: the number of projects of new products development
S: the number of total projects of technical development
To standardize the original date X_{ij} in Table 4.1 by

$$x_{ij}=X_{ij}/ \max X_{ij} \tag{5}$$

for all $i=1,2,\ldots\ldots,10; j=1,2,\ldots\ldots,12$.

We have the standardized original data matrix (See Table 4).

(Arrange attributes as the same order as Table 3)

We assume that 4 experts are invited to participate in the evaluation of the priorities in the beginning（full mark is 5）. The original priority matrix is given as follows (See Table 5).

Table 5. The original priority matrix given by the experts in the first instance

1.0	1.5	2.0	2.0	3.0	1.0	3.0	5.0	2.0	1.0	0.5	2.0
1.5	2.0	2.0	2.0	4.0	1.5	4.0	4.0	1.0	2.0	1.0	3.0
1.2	1.0	3.0	2.5	3.5	1.2	3.0	4.0	1.5	2.0	1.0	3.0
1.0	2.0	3.0	2.0	3.5	1.5	3.0	4.0	1.5	1.0	0.5	2.0

So, we have the normalized original priority matrix (See Table 6).
The bounds of values of 12 priorities of attributes:
The lower limit α_j $(j = 1,2,\cdots\cdots,12)$: (Lower bound vector (or matrix))

(0.04000 0.03717 0.07143 0.07143 0.12500 0.04167 0.11152 0.14286 0.03571 0.04000 0.02000 0.08000)

The upper limit β_j $(j = 1,2,\cdots\cdots,12)$: (Upper bound vector (or matrix))

(0.05357 0.08000 0.12000 0.09294 0.14286 0.06000 0.14286 0.20833 0.08333 0.07435 0.03717 0.11152)

Table 6. The normalized original priority matrix

0.04167	0.06250	0.08333	0.08333	0.12500	0.04167	0.12500	0.20833	0.08333	0.04167	0.02083	0.08333
0.05357	0.07143	0.07143	0.07143	0.14286	0.05357	0.14286	0.14286	0.03571	0.07143	0.03571	0.10714
0.04461	0.03717	0.11152	0.09294	0.13011	0.04461	0.11152	0.14870	0.05576	0.07435	0.03717	0.11152
0.04000	0.08000	0.12000	0.08000	0.14000	0.06000	0.12000	0.16000	0.06000	0.04000	0.02000	0.08000

Table 7. The synthetic priorities get from 1st to 4th repeat

1	0.0453	0.0701	0.0827	0.0899	0.1286	0.0581	0.1409	0.1564	0.0508	0.0418	0.0239	0.1115
2	0.0468	0.0772	0.0746	0.0927	0.1250	0.0600	0.1429	0.1636	0.0451	0.0401	0.0205	0.1115
3	0.0467	0.0777	0.0737	0.0929	0.1250	0.0600	0.1429	0.1664	0.0432	0.0400	0.0200	0.1115
4	0.0469	0.0777	0.0735	0.0929	0.1250	0.0600	0.1429	0.1664	0.0432	0.0400	0.0200	0.1115

The synthesized original priority vector:

(0.00450 0.0627 0.0967 0.0819 0.1346 0.0500 0.1248 0.1644 0.0584 0.0571 0.0286 0.0957)

Here ξ=1.000E-03.

According (4), we have the synthesized priorities get from the repeats (See Table 7). After 4th repeat:

ξ1=3.603E-02, ξ2=1.568E-02, ξ3=3.537E-03, ξ4=1.890E-04<1.000E-03.

And the values of objective functions are as follows:

Beijing	Shanghai	Jiangsu	Zhejiang	Anhui	Fujian	Jiangxi	Shandong	Henan	Guangdong
0.6564	0.8339	0.8927	0.8272	0.5923	0.6057	0.4901	0.7532	0.5435	0.7082

So, we have the ability ratings of the units:

1	2	3	4	5	6	7	8	9	10
Jiangsu	Shanghai	Zhejiang	Shandong	Guangdong	Beijing	Fujian	Anhui	Henan	Jiangxi

7 Summary

Formerly, an important problem is ignored while the group decision methods are applied in evaluations. The problem is that the preferences or attitudes of evaluated units are not under considering. This paper proposes a new approach to solve the priorities of the attributes while the preferences of evaluated units are under considering. The method uses an interactive method to optimize the priorities of attributes and determine the scores of evaluated units by simulating the decision-making process. And the process of this method is showed by an applied example given at the end of the paper.

By the way, the proposed method can be deeply applied in interactive economic and operation system. For example, this method can be utilized for calculating priorities of companies in supply chain performance evaluation.

References

1. Satty, T.L.: The Analytic Hierarchy Process. McGraw-Hill, New York (1980)
2. Saaty, T.L.: The Analytic Hierarchy and Analytic Network Measurement Processes: Applications to Decisions under Risk. European Journal of Pure and Applied Mathematics 1, 122–196 (2008)
3. Forman, E.H., Selly, M.A.: Decision By Objectives. World Scientific, Singapore (2001)
4. Ozdemir, M.S., Saaty, T.L.: The unknown in decision making: What to do about it. European Journal of Operational Research 174, 349–359 (2006)
5. Vargas, L.G.: An overview of the analytic hierarchy process and its applications. European Journal of Operation Research 48, 2–8 (1990)
6. Acezel, J., Satty, T.L.: Procedures for synthesizing ratio judgments. Journal of Mathematical Psychology 27, 93–102 (1983)

7. Bryson, N.: A goal programming method for generating priority vectors. Journal of Operational Research 46, 641–648 (1995)
8. Satty, T.L.: Axiomatic foundation of the analytical hierarchy process. Management Science 32, 841–855 (1986)
9. Anderson, T.W.: An Introduction to Multivariate Statistical Analysis, 2nd edn. John Wiley & Sons, Inc., New York (1984)
10. Xiaojian, C., Liang, L.: Methods and applications of system evaluation. University of Science and Technology of China Press, Hefei (1993)

Solving Delay Differential Equations with Homotopy Analysis Method

Qi Wang[1,3,*] and Fenglian Fu[2]

[1] Faculty of Applied Mathematics, Guangdong University of Technology,
Guangzhou 510006, China
bmwzwq@126.com
[2] School of Environmental Science and Engineering, Guangdong University of Technology,
Guangzhou 510006, China
[3] Shenzhen Graduate School, Harbin Institute of Technology, Shenzhen 518055, China

Abstract. Delay differential equations have a wide range of application in science and engineering. They arise when the rate of change of a time-dependent process in its mathematical modeling is not only determined by its present state but also by a certain past state. In this paper, a nonlinear delay differential equation in biology was investigated. The approximation solution for the model was obtained by homotopy analysis method. Different from other analytic techniques, the homotopy analysis method provides a simple way to ensure the convergence of the solution series, so that one can always get accurate approximations. Compared with the numerical solution, the approximation solution has higher precision. It is showed that the homotopy analysis method was valid and feasible to the study of delay differential equations.

Keywords: Delay differential equations; Analytic solutions; Homotopy analysis method.

1 Introduction

Delay differential equations (DDEs) provide a powerful model of many phenomena in applied sciences. Some studies in biology, economy, control and electrodynamics etc. [1-3] have shown that DDEs play an important role in explaining many different phenomena. Particularly, DDEs turn out to be fundamental when the model based on ordinary differential equations fails.

Several models with delay were analyzed by Hethcote [4] and Cooke and Kaplan [5]. Singh [6] has discussed some delay epidemic models qualitatively. And Berezansky and Braverman [7] established some oscillation and non-oscillation conditions for nonlinear equations arising in population dynamics. To the best of our knowledge, DDEs are handled by numerical techniques [8-10] mostly. It is very difficult even impossible to obtain the analytic solution.

---

* Correspondig author.

K. Li et al. (Eds.): LSMS / ICSEE 2010, Part I, CCIS 97, pp. 144–153, 2010.
© Springer-Verlag Berlin Heidelberg 2010

In order to solve practical problems, different perturbation techniques have been used in engineering [11,12]. Mostly, these perturbation techniques lead to some important results, but they can not be applied to all nonlinear problems. In recent years, Liao [13-16] has proposed a nonlinear analytic technique, named the homotopy analysis method (HAM). The HAM does not require small parameters and can be used to solve different nonlinear problems [17-20]. However, the application of HAM in DDEs is less in recent years.

In this paper we use HAM to investigate a delay logistic equation in biological systems

$$\frac{dx(t)}{dt} = rx(t)\left[1 - \frac{x(t-\tau)}{k}\right], \tag{1}$$

where x(t) is the amount of biology species at time t, the nonnegative parameters r and k are the intrinsic growth rate and the environmental carrying capacity, respectively, and τ is the time-delay. In this article, we obtained approximate analytic solution of (1) by HAM. Different from [21,22], our process is more easier to implement.

2 Model Description and the HAM

We consider the delay model

$$\begin{cases} \dfrac{dx(t)}{dt} = rx(t)\left[1 - \dfrac{x(t-\tau)}{k}\right], \\ x(0) = x_0, \end{cases} \tag{2}$$

here x_0 is the initial amount.

Let $\mu = \dfrac{r}{k}$, we have

$$\begin{cases} \dfrac{dx(t)}{dt} - rx(t) + \mu x(t)x(t-\tau) = 0, \\ x(0) = x_0. \end{cases} \tag{3}$$

In the next part, we extend Liao's ideas to DDEs by applying the HAM to (3).

Let us consider the equation

$$DD(x(t)) = 0, \tag{4}$$

where DD is a delay differential operator, t denotes independent variable and $x(t)$ is an unknown function. For simplicity, we ignore initial condition, which can be treated in the same way.

The following zero-order deformation equation is

$$(1-q)L[u(t,q) - x_0(t)] = qh\,DD(u(t,q)), \tag{5}$$

where $q \in [0,1]$ is the embedding parameter, h is a nonzero auxiliary parameter, L is an auxiliary linear operator possessing the property $L(C) = 0$, $x_0(t)$ is an initial guess of $x(t)$, $u(t,q)$ is an unknown function on independent variables t,q. One has great freedom to choose auxiliary parameter h in HAM. For $q = 0$ and $q = 1$, it holds

$$\begin{cases} u(t,0) = x_0(t), \\ u(t,1) = x(t). \end{cases} \tag{6}$$

Thus as q increases from 0 to 1, the solution $u(t,q)$ varies from the initial guess $x_0(t)$ to the solution $x(t)$. Expanding $u(t,q)$ in Taylor series with respect to q, one has

$$u(t,q) = x_0(t) + \sum_{m=1}^{+\infty} x_m(t)q^m, \tag{7}$$

where

$$x_m(t) = \frac{1}{m!} \frac{\partial^m u(t,q)}{\partial q^m}\bigg|_{q=0}. \tag{8}$$

If the auxiliary linear operator, the initial guess, and the auxiliary parameter h are so properly chosen, the series (7) converges at $q = 1$, one has

$$x(t) = x_0(t) + \sum_{m=1}^{+\infty} x_m(t). \tag{9}$$

For similarity, we define the vector

$$\vec{x}_m(t) = [x_0(t), x_1(t), \cdots, x_m(t)]. \tag{10}$$

Differentiating (5) m times with respect to q and then setting $q = 0$ and finally dividing them by $m!$, we have the so-called m th-order deformation equation

$$L[x_m(t) - X_m x_{m-1}(t)] = h\,\mathrm{NDR}(\vec{x}_{m-1}(t)), \tag{11}$$

where

$$\mathrm{NDR}(\vec{x}_{m-1}(t)) = \frac{1}{(m-1)!} \frac{\partial^{m-1} \mathrm{DD}(u(t,q))}{\partial q^{m-1}}\bigg|_{q=0}, \tag{12}$$

and

$$X_m = \begin{cases} 0, & m \le 1, \\ 1, & m > 1. \end{cases} \tag{13}$$

The m th-order deformation equation (11) is linear and hence can be easily solved, especially by means of symbolic computation software such as Mathematica, Maple and so on.

3   Analytic Solution

To demonstrate the effectiveness of the method, we consider two different auxiliary linear operators. For simplicity, we assume that the constant of integration $C = 0$.

3.1   Auxiliary Linear Operator $L_1[u(t,q)] = \dfrac{\partial u(t,q)}{\partial t}$

We choose the initial guess

$$x_0(t) = x_0 e^{rt}. \tag{14}$$

Furthermore, (3) suggests to defining the nonlinear differential operator

$$NDD[u(t,q)] = \frac{\partial u(t,q)}{\partial t} - ru(t,q) + \mu u(t,q)u(t-\tau,q). \tag{15}$$

Using above definition, we construct the zeroth-order deformation equation

$$(1-q)L[u(t,q) - x_0(t)] = qh DD(u(t,q)). \tag{16}$$

Obviously, when q = 0 and q = 1,

$$u(t,0) = x_0(t), \quad u(t,1) = x(t). \tag{17}$$

According to (11)–(13), we gain the m th-order deformation equation

$$L[x_m(t) - X_m x_{m-1}(t)] = h N DR(\vec{x}_{m-1}(t)), \tag{18}$$

where

$$N DR[\vec{x}_{m-1}(t)] = x'_{m-1}(t) - rx_{m-1}(t) + \mu \sum_{j=0}^{m-1} x_j(t)x_{m-1-j}(t-\tau). \tag{19}$$

Now, the solution of (18) for $m \geq 1$ becomes

$$x_m(t) = X_m x_{m-1}(t) + hL^{-1}[N DR(\vec{x}_{m-1}(t))]. \tag{20}$$

From (14) , (17), (19) and (20), we now successively obtain

$$x_1(t) = \frac{1}{2r} h\mu x_0^2 e^{-r\tau} e^{2rt},$$

$$x_2(t) = \frac{1}{4r} h\mu x_0^2 e^{-r\tau}(h+2)e^{2rt} + \frac{1}{6r^2} h^2 \mu^2 x_0^3 (e^{-2r\tau} + e^{-3r\tau})e^{3rt}, \tag{21}$$

$$x_3(t) = \frac{1}{8r} h\mu x_0^2 e^{-r\tau}(h+2)^2 e^{2rt} + \frac{1}{36r^2} h^2 \mu^2 x_0^3 (e^{-2r\tau} + e^{-3r\tau})(7h+12)e^{3rt}$$

$$+ \frac{1}{48r^3} h^3 \mu^3 x_0^4 (2e^{-3r\tau} + 5e^{-4r\tau} + 2e^{-5r\tau} + 2e^{-6r\tau})e^{4rt}. \tag{22}$$

The solution of (2) in series form is given by

$$x(t) = x_0(t) + x_1(t) + x_2(t) + x_3(t) + \cdots. \tag{23}$$

3.2 Auxiliary Linear Operator $L_2[u(t,q)] = \dfrac{\partial u(t,q)}{\partial t} - ru(t,q)$

Similar to Section 3.1, we have

$$x_0(t) = x_0 e^{rt},$$

$$x_1(t) = \frac{1}{r} h\mu x_0^2 e^{-r\tau} e^{2rt}, \tag{24}$$

$$x_2(t) = \frac{1}{r} h\mu x_0^2 e^{-r\tau} (h+1) e^{2rt} + \frac{1}{2r^2} h^2 \mu^2 x_0^2 (e^{-2r\tau} + e^{-3r\tau}) e^{3rt},$$

$$x_3(t) = \frac{1}{r} h\mu x_0^2 e^{-r\tau} (h+1)^2 e^{2rt} + \frac{1}{2r^2} h^2 \mu^2 x_0^2 (e^{-2r\tau} + e^{-3r\tau})[1 + (2h+1)x_0] e^{3rt}$$
$$+ \frac{1}{6r^3} h^3 \mu^3 x_0^3 [e^{-3r\tau} + (1+2x_0) e^{-4r\tau} + e^{-5r\tau} + e^{-6r\tau}] e^{4rt}. \tag{25}$$

The solution of (2) in series form is given by

$$x(t) = x_0(t) + x_1(t) + x_2(t) + x_3(t) + \cdots. \tag{26}$$

4 Discussion

The series (23) and (26) contain the auxiliary parameter h. Once one of them is convergent, it must be the exact solution of (2). The validity of the method is based on such an assumption that the series (23) and (26) converge at $q = 1$. It is the auxiliary parameter h which ensures that this assumption can be satisfied. As pointed out by Liao [14], in general, by means of the so-called h-curves, it is straightforward to choose a proper value of h which ensures that the solution series is convergent. So the auxiliary parameter h provides us with a flexible and convenient way to adjust and control convergence region and rate of solution series. Fig. 1 describes the 2nd order and 3rd order approximation solution of (2) by the HAM with $x_0 = 1$, $r = 0.3$, $k = 200$ and $\tau = 0.5$ at $t = 0$.

From Fig. 2, we can see that the time-delay τ has indeed a great influence on the global dynamics of the system. Using explicit Runge-Kutta method, we can obtain numerical solution. Compared with the 3rd order approximation solution, the latter has higher precision (as shown in Figs. 5-6).

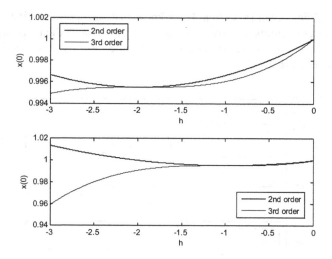

Fig. 1. The so-called h-curve of $x(0)$ when $x_0 = 1$, $r = 0.2$, $k = 200$ and $\tau = 0.5$. (Upper: (21) and (22) with L_1, Lower: (24) and (25) with L_2)

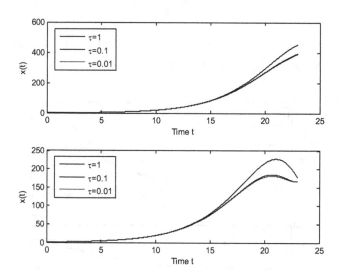

Fig. 2. Investigation of the 3rd order approximate solution of $x(t)$ with different τ when $x_0 = 1$, $r = 0.3$, $k = 200$ and $h = -0.15$. (Upper: operator L_1, Lower: operator L_2)

We can also see that the linear operators L_1 and L_2 are both effective from all the figures. So one has great freedom to choose auxiliary linear operator. But how to select the better operator and the optimal one needs further research.

From the series given in section 3, we know that the approximation solution by HAM is consistent with the one by Adomian decomposition method (ADM) or homotopy perturbation method (HPM) at $h = -1$. In other words, ADM and HPM are special cases of HAM.

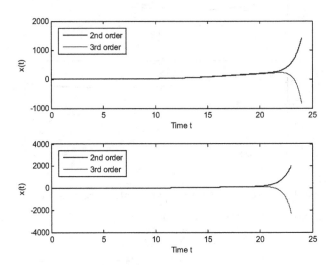

Fig. 3. The different order approximation of $x(t)$ with $x_0 = 1$, $r = 0.3$, $k = 200$, $\tau = 0.5$ and $h = -0.5$. (Upper: operator L_1, Lower: operator L_2)

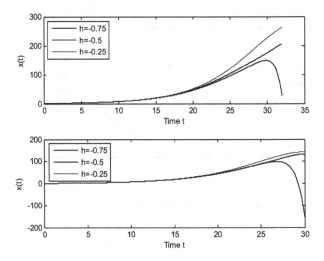

Fig. 4. Investigation of the 3rd order approximate solution of $x(t)$ with different h when $x_0 = 1$, $r = 0.2$, $k = 200$ and $\tau = 0.5$. (Upper: operator L_1, Lower: operator L_2)

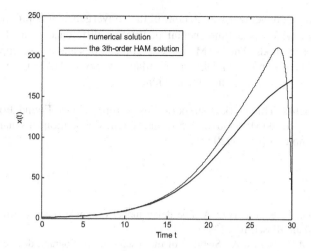

Fig. 5. Comparison of the numerical solution with the 3rd order approximate solution of $x(t)$ with operator L_1, where $x_0 = 1$, $r = 0.23$, $k = 200$, $\tau = 0.5$ and $h = -0.5$

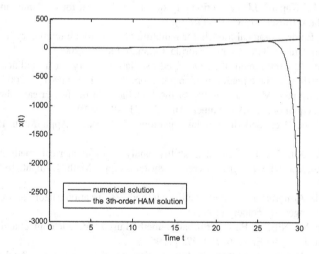

Fig. 6. Comparison of the numerical solution with the 3rd order approximate solution of $x(t)$ with operator L_2, where $x_0 = 1$, $r = 0.23$, $k = 200$, $\tau = 0.5$ and $h = -0.5$

5 Conclusions

In this paper, the HAM is successfully applied to solving nonlinear DDEs in biology. Based on the symbolic computation software such as Mathematica or Maple, the HAM is directly extended to derive explicit solution of the DDEs. And the HAM provides us with a convenient way to control the convergence of approximation series by adapting h, which is a fundamental qualitative difference in analysis between HAM and other methods. The HAM has general meanings and can be applied to solve some other types of nonlinear DDEs in a similar way. So the HAM is a very powerful method with large potential in nonlinear science.

Acknowledgments. This work is supported by Nature Science Foundation of Guangdong Province (No. 9451009001002753) and Youth Foundation of Guangdong University of Technology (No. 092027).

References

1. Aiello, W.G., Freedman, H.I.: A time delay model of single species growth with stage structure. Math. Biosci. 101, 139–156 (1990)
2. Buhmann, M.D., Iserles, A.: Stability of the discretized pantograph differential equation. Math. Comput. 60, 575–589 (1993)
3. Ockendon, J.R., Tayler, A.B.: The dynamics of a current collection system for an electric locomotive. Proc. R. Soc. London Ser. A 322, 447–468 (1971)
4. Hethcote, H.W.: The mathematics of infectious diseases. SIAM Rev. 42, 599–653 (2000)
5. Cooke, K.L., Kaplan, J.L.: A periodicity threshold theorem for epidemic and population growth. Math. Biosci. 31, 87–104 (1976)
6. Singh, N.: Epidemiological models for mutating pathogen with temporary immunity. PhD dissertation, University of Central Florida, Orlando, Florida (2006)
7. Berezansky, L., Braverman, E.: Linearized oscillation theory for a nonlinear nonautonomous delay differential equation. J. Comput. Appl. Math. 151, 119–127 (2003)
8. Liu, M.Z., Spijker, M.: The stability of the θ methods in the numerical solution of delay differential equations. IMA J. Numer. Anal. 10, 31–48 (1990)
9. Baker, C.T.H.: Retarded differential equations. J. Comput. Appl. Math. 125, 309–335 (2000)
10. Zhao, J.J., Xu, Y., Liu, M.Z.: Stability analysis of numerical methods for linear neutral Volterra delay-integro-differential system. Appl. Math. Comput. 167, 1062–1079 (2005)
11. Rand, R.H., Armbruster, D.: Perturbation Methods, Bifurcation Theory and Computer Algebraic. Springer, Heidelberg (1987)
12. Andersson, M., Nilsson, F.: A perturbation method used for static contact and low velocity impact. Int. J. Impact Eng. 16, 759–775 (1995)
13. Liao, S.J.: Beyond Perturbation: Introduction to the Homotopy Analysis Method. Chapman and Hall/ CRC Press, Boca Raton (2003)
14. Liao, S.J.: An approximate solution technique which does not depend upon small parameters: a special example. Int. J. Non-linear Mech. 30(3), 371–380 (1995)
15. Liao, S.J.: On the homotopy analysis method for nonlinear problems. Appl. Math. Comput. 147(2), 499–513 (2004)
16. Liao, S.J., Tan, Y.: A general approach to obtain series solutions of nonlinear differential equations. Stud. Appl. Math. 119, 297–340 (2007)
17. Abbasbandy, S.: Soliton solutions for the Fitzhugh–Nagumo equation with the homotopy analysis method. Appl. Math. Model. 32, 2706–2714 (2008)

18. Jafari, H., Seifi, S.: Homotopy analysis method for solving linear and nonlinear fractional diffusion-wave equation. Commun. Nonlinear Sci. Numer. Simulat. 14, 2006–2012 (2009)
19. Aliakbar, V., Alizadeh-Pahlavan, A., Sadeghy, K.: The influence of thermal radiation on MHD flow of Maxwellian fluids above stretching sheets. Commun. Nonlinear Sci. Numer. Simulat. 14, 779–794 (2009)
20. Odibat, Z., Momani, S., Xu, H.: A reliable algorithm of homotopy analysis method for solving nonlinear fractional differential equations. Appl. Math. Model. 34, 593–600 (2010)
21. Wu, Z.K.: Solution of the ENSO delayed oscillator with homotopy analysis method. J. Hydrodynamics. 21, 131–135 (2009)
22. Khan, H., Liao, S.J., Mohapatra, R.N., Vajravelu, K.: An analytical solution for a nonlinear time-delay model in biology. Commun. Nonlinear Sci. Numer. Simulat. 14, 3141–3148 (2009)

EEG Classification Based on Artificial Neural Network in Brain Computer Interface

Ting Wu[1], Banghua Yang[2], and Hong Sun[3]

[1] School of Mechanical Engineering, Shanghai Dianji University, 200240, Shanghai, China
[2] School of Mechatronics Engineering & Automation,
Shanghai University, 200072, Shanghai, China
[3] School of Mechanical and Electrical Engineering, Anhui University of Architecture,
230000, Hefei, China
wut@sdju.edu.cn

Abstract. Aiming at the topic of electroencephalogram (EEG) pattern recognition in brain computer interface (BCI), a classification method based on probabilistic neural network (PNN) with supervised learning was presented in this paper. It applied the recognition rate of training samples to the learning progress of network parameters, The learning vector quantization is employed to group training samples and the Genetic algorithm (GA) is used for training the network's smoothing parameters and hidden central vector for determining hidden neurons. Utilizing the standard dataset I(a) of BCI Competition 2003 and comparing with other classification methods, the experiment results show that this way has the best performance of pattern recognition, and the classification accuracy can reach 93.8%, which improves over 5% compared with the best result (88.7%) of the competition. This technology provides an effective way to EEG classification in practical system of BCI.

Keywords: Probabilistic neural network (PNN), classification, brain computer interface (BCI), electroencephalogram (EEG).

1 Introduction

A brain-computer interface (BCI) is an alternative communication and control channel that does not depend on the brain's normal output pathway of peripheral nerves and muscles [1]. A BCI system is intended to help severely disabled people to communicate with computers or control electronic devices through their thoughts. The most common BCI systems are based on the analysis of spontaneous electroencephalogram (EEG) signals. Over the last years evidence has accumulated to show the possibility to recognize a few mental tasks from spontaneous EEG signals [2~5]. Despite some successful demonstrations of BCI, there are a number of issues need to be solved before this technology can move outside the lab, especially to fulfill the hopes of disabled people. One of the issues is how to effectively recognize distinguishable EEG signals and so to differentiate human intentions.

K. Li et al. (Eds.): LSMS / ICSEE 2010, Part I, CCIS 97, pp. 154–162, 2010.

The EEG recognition procedure mainly includes the feature extraction, feature selection and the classification, in which the classification is the object of BCI systems [6]. This paper mainly focuses on the classification. At present, classification methods for spontaneous EEG mainly include such as linear classifier [7], Kalman filter [8], ANN (artificial neural network)[9], and etc. Linear classifier is simple and realized easily, it needs less computation time and storage capacity. However, since EEG is nonlinear, and the feature vector is non-linearly separable, this kind of classifier has not a high precision of recognition. Kalman filter is an algorithm based on probability theory, it needs transcendent knowledge which is difficult to be obtained since EEG is a very complicated physiological signal. ANN is a comparatively good machine learning method. It can solve many nonlinear problems and has been successfully applied to a wide variety of engineering problems. In this paper, we propose an effective classification method based on probabilistic neural network with supervised learning. The learning vector quantization is employed to group training samples and the genetic algorithm(GA) is used for training the network's smoothing parameters and hidden central vector for determining hidden neurons. Utilizing the standard dataset of BCI Competition 2003, several kinds of classification method are compared, and the results demonstrate that this technology can provide better performances in classification efficiency of EEG signal.

2   The Structure of Probability Neural Network and Algorithm Description

The neural network structure can be illustrated in Fig.1. Four layers, one input layer, one output layer, one hidden layer and one added layer, are designed [10]. The input layer transfers $\vec{x}$ to the network without any computation. After the hidden layer receiving $\vec{x}$, the output and input of the j_{th} neuron at the i_{th} mode is defined by equation (1),

$$\phi_{ij}(\vec{x}) = \frac{1}{(2\pi)^{d/2}\sigma^d} \exp\left[-\frac{(\vec{x}-\vec{x_{ij}})(\vec{x}-\vec{x_{ij}})^T}{\sigma^2}\right] \tag{1}$$

where $i = 1,2,\cdots,M$, $j = 1,2,\cdots,N_i$, M is the number of regimentations of training samples, N_i is the amount of training samples that belong to i_{th} class, also defined as the number of neurons at the i_{th} mode in hidden layer. d is the dimension of sample, $\sigma \in (0, \infty)$ is smoothing parameter, $\overrightarrow{x_{ij}}$ is the j_{th} training

sample of the i_{th} class and called as the j_{th} hidden central vector of PNN. The func-

tion of the added layer is illustrated by equation (2), the outputs of the neurons in

hidden layer whose hidden central vectors belong to the same class are added and

averaged.

$$f_{iN_i}(\overrightarrow{x}) = \frac{1}{N_i} \sum_{j=1}^{N_i} \phi_{ij}(\overrightarrow{x}) \qquad (2)$$

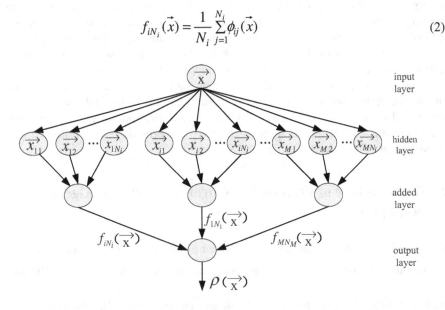

Fig. 1. Structure of probabilistic neural network (PNN)

According to Bayesian decision, if every kind of prior probability is equal and the

loss function is defined as 0/1 loss, i is the class evaluation of the training sample $\overrightarrow{x}$

while $f_{iN_i}(\overrightarrow{x})$ reaches the maximum, i.e., is the output of the whole network $\rho(\overrightarrow{x})$.

$$\rho(\overrightarrow{x}) = \arg\max\{f_{iN_i}(\overrightarrow{x})\} \qquad (3)$$

Here, *arg* is the operation of getting the value of subscript i in $f_{iN_i}(\overrightarrow{x})$.

3 Supervised Learning Based on Genetic Algorithm

The total number of training samples is $N = \sum\limits_{i=1}^{M} N_i$, if n samples were input into the

network and could get the accurate results of classification, $f = n/N$ represents the

classification accuracy. $v_{ij}(i=1,2,\cdots,M; j=1,2,\cdots,C_i)$ are clustering points of

training samples, C_i is the clustering number of the i_{th} mode, the center of this

mode is $\overrightarrow{m_i} = \dfrac{\sum\limits_{j=1}^{N_i} \overrightarrow{x_{ij}}}{N_i}$, in the i_{th} pattern, the nearest clustering point to $\overrightarrow{m_i}$ is $\overrightarrow{v_{is_i}}$,

i.e., $s_i = \arg\min\limits_{j}\left\|\overrightarrow{v_{ij}} - \overrightarrow{m_i}\right\|(i=1,2,\cdots,M)$, $\|\cdot\|$ represents the Euclidean distance,

if $v_{ij}(i=1,2,\cdots,M; j\neq s_i, j=1,2,\cdots,C_i)$ is directly defined as the j_{th} hidden

center vector of the i_{th} pattern, $\lambda_{is_i}(i=1,2,\cdots,M)$ and σ are changeable, here,

$\lambda_{is_i}(i=1,2,\cdots,M)$ represents the s_i hidden center vector of the i_{th} pattern, the

classification accuracy (f) is only determined by σ and $\lambda_{is_i}(i=1,2,\cdots,M)$ as

defined in equation (4).

$$f = F(\sigma, \lambda_{1s_1}, \lambda_{2s_2}, \cdots, \lambda_{Ms_M}) \tag{4}$$

So, it needs selecting appropriate σ and λ_{is_i} to reach the highest classification

accuracy.

Assuming an interval $\sigma \in [0,\theta_1]$, $\lambda_{is_k} \in [v_{is_ik} - \theta_2 d_{i\max}, v_{is_ik} + \theta_2 d_{i\max}]$

$(k=1,2,\cdots,d)$, where $d_{i\max}$ is the maximal Euclidean distance of the i_{th} class in training samples, i.e., $d_{i\max} = \max\limits_{s,j,s\neq j}\{\|\overrightarrow{x_{is}} - \overrightarrow{x_{ij}}\|\}$. θ_1 and θ_2 are determinate as

constants ($\theta_1 > 0$, $0 < \theta_2 < 1$), λ_{is_ik} and v_{is_ik} represent the k_{th} dimension value of

$\overrightarrow{\lambda_{is_i}}$ and $\overrightarrow{v_{is_i}}$ respectively. With fixed training datasets, the above-mentioned opti-

mization problem can be defined as following.

$$\max\{N, f\}$$
$$\sigma \in [0, \theta_1] \qquad\qquad (5)$$
$$\lambda_{is_ik} \in [v_{is_ik} - \theta_2 d_{i\max}, v_{is_ik} + \theta_2 d_{i\max}], i = 1, 2, \cdots, M; k = 1, 2, \cdots, d$$

How to select the smoothing parameter and make classification accuracy reach the

optima in theory is a hot research topic [11]. In this paper, it selects $\sigma \in [0, \theta_1]$.

Since the s_i hidden center vector of the i_{th} pattern should be located around

the   clustering   points   of s_i in   the i_{th} class,   it   determines

$$\lambda_{is_k} \in [v_{is_ik} - \theta_2 d_{i\max}, v_{is_ik} + \theta_2 d_{i\max}].$$

Genetic algorithm (GA) is a random searching and optimization method. There are
many merits with GA, such as better versatility, simple searching process, high capability of global searching and etc. So this paper used GA to solve the above optimization problem.

The number of clustering-- $C_i (i = 1, 2, \cdots, M)$ must be considered in this algo-

rithm. It corresponds to the number of neurons in hidden layer. There is no theoretic
rule of selecting the number of clustering, generally, an experiential formula usually

be utilized, $C_i \le \sqrt{N_i}$, C_i often be determined as a material value in advance [12].

Since the selected area is close, the classification accuracy (f) must have a max-

imum in the area. By adopting the elitist preservation strategy, the algorithm can converge to the optimal solution which takes the value "1" with probability [13].

4   Experiment and Classification Result

4.1   Dataset

All data were acquired from a single healthy subject at the University of Tuebingen, Germany, as described in [14]. As shown in Fig.2, six EEG electrodes were all referenced to the vertex electrode C_z (International 10-20 system) as follows: channel 1 and 2, left and right mastoids; channel 3-6, anterior (ch.3, ch.5) or posterior (ch.4, ch.6) to position C3 (ch.3, ch.4) or C4 (ch.5, ch.6). These six EEG voltages were sampled at 256Hz.

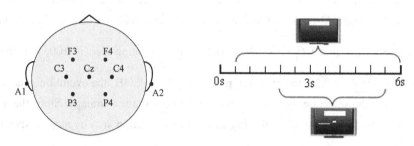

Fig. 2. Map of EEG electrodes for 6 channels   **Fig. 3.** Experiment paradigm

Trials consisted of three phases: a 1-s rest phase, 1.5-s cue-presentation phase, and a 3.5-s feedback phase as shown in Fig.3. At the beginning of the 1.5-s cue-presentation phase, a visual target indicator appeared either at the top ("cue 0" trials, instructing the subject to strive for cortical negativity, defined below) or bottom ("cue 1" trials, cortical positivity) of the screen. The target remained the current level of cortical negativity being generated by the subject.

Cortical negativity was defined at the running average of the voltages on the two mastoid electrodes (channels 1 and 2) over the past 0.5s, relative to the cue-presentation phase. Because these electrodes are referenced to C_z, positive values correspond to cortical negativity. Cortical negativity is a form of SCP which has been associated with a wide range of behavioral states pertaining to altering, anticipation, and preparation.

The trials were separated into a training set (268 trials) and a test set (293 trials), both of which contained EEG data from only the feedback phase of each trial. The cue labels (class "cue 0" or "cue 1") for the training set was used to tune the parameters of the classification algorithm, whose performance was subsequently assessed on the test set.

4.2 Result of EEG Classification

Dealing with the standard data of EEG, the first step is feature extraction and get a feature vector of P dimensions, the features of training samples are acted as the input of PNN, and the classes of thinking are acted as the output. The features include the means in time domain of six channels (6 dimensions), the means of power spectrum (6 dimensions) and the means of time (6 dimensions), so $P = 18$. There are 268 groups of training samples, 135 groups belong to class "0", 133 groups belong to class "1", and 293 groups of testing samples. The number of neurons in hidden layer of each class is 10, $\theta_1 = 0.5, \theta_2 = 0.5$, the population size is 100, the crossover probability $P_c = 0.55$, the mutation probability $P_m = 0.001$, the evolution generation is 800, and the selection operation adopts elitist preservation strategy. Since the average fitness function approaches the maximal fitness function step by step, PNN must be convergent.

Table 1. Results of different classification methods

algorithm	accuracy of training samples	accuracy of testing samples
Fisher	85.1%	75.3%
BP	91.2%	86.8%
RBF	93.6%	89.5%
FCM-PNN	96.5%	91.6%
PNN(NS)	97.7%	91.3%
GA-PNN(S)	99.3%	93.8%

In order to prove the performance of this technology, the evaluation indicator is the classification accuracy of test samples that is averaged by 10 runs, the results of different classification methods are shown as Table.1. Fisher is a linear classifier, BP is back propagation neural network, RBF is radial basis function neural network, FCM-PNN is PNN combining with fuzzy c-mean, PNN (NS) is only PNN without supervised learning, GA-PNN(S) is PNN based on genetic algorithm with supervised learning. From Table.1, GA-PNN(S) has reached the maximal classification accuracy.

5 Conclusion

A novel classification method based on probabilistic neural network (PNN) with supervised learning was proposed in this paper, in which the learning vector quantization is employed to group training samples and the Genetic algorithm(GA) is used for training the network's smoothing parameters and hidden central vector for determining hidden neurons. Using standard datasets, the proposed method was compared with linear classifier, other neural networks and the algorithm without supervised learning. Simulation results show that this method provides the best recognition performance and is very applicable for EEG classification in brain computer interface.

Acknowledgments

This paper is supported by Leading Academic Discipline Project of Shanghai Municipal Education Commission (No. J51902), Shanghai Education Commission Foundation for Excellent Youth High Education Teacher (No. Sdj08001) and Shanghai "Chen Guang" Project (No.09CG69).

References

1. Wolpaw, J.R., Birbaumer, N., Heetderks, W.J.: Brain computer interface technology: a review of the first international meeting. IEEE Trans. Rehab. Eng. 8(2), 64–73 (2000)
2. Millán, J., Mouriño, J., Franzé, M., et al.: A local neural classifier for the recognition of EEG patterns associated to mental tasks. IEEE Trans. Neural Networks 13(3), 678–686 (2002)
3. Millán, J., Mouriño, J.: Asynchronous BCI and local neural classifiers: an overview of the adaptive brain interface project. IEEE Trans. Neural Syst. Rehab. Eng. 11(2), 159–161 (2003)
4. Wolpaw, J.R., McFarland, D.J., Neat, G.W., et al.: An EEG-based brain–computer interface for cursor control. Electroenceph Clin. Neurophysiol. 78(3), 25–29 (1991)
5. Birbaumer, N., Ghanayim, N., Hinterberger, T., et al.: A spelling device for the paralyzed. Nature 398(5), 297–298 (1999)
6. Pfurtscheller, G., Neuper, C., Schlogl, A., et al.: Separability of EEG signals recorded during right and left motor imagery using adaptive autoregressive parameters. IEEE Trans. Rehab. Eng. 6(3), 316–325 (1998)
7. Matthias, K., Peter, M.: BCI competition 2003-Data set IIb: support vector machines for the P300 speller paradigm. IEEE Transactions on Biomedical Engineering 51(6), 1073–1076 (2004)
8. Deriche, M., Al-Ani, A.: A new algorithm for EEG feature classification using mutual information. In: Proceedings of the 2001 IEEE International Conference on Acoustics, Speech, and Signal Processing, Salt Lake City, pp. 1057–1060 (2001)
9. Petersa, B.O., Pfurtschellerb, G., Flyvbjerg, H.: Mining multi-channel EEG for its information content: an ANN-based method for a brain-computer interface. Neural Networks 11(2), 1429–1433 (1998)
10. Specht, D.F.: Probabilistic neural network. Neural Network 3(2), 109–118 (1998)

11. Rutkowski, L.: Adaptive probabilistic neural networks for pattern classification in time-varying environment. IEEE Trans. on Neural Network 15(4), 811–827 (2004)
12. Hoya, T.: On the capability of accommodating new classes within probabilistic neural networks. IEEE Trans. on Neural Network 14(2), 450–453 (2003)
13. Rudolph, G.: Convergence analysis of canonical genetic algorithms. IEEE Transactions on Neural Networks 5(1), 96–101 (1994)
14. Brett, D.M., Justin, W., Seung, H.S.: BCI Competition 2003-Data Set Ia: Combining gamma-band power with slow cortical potentials to improve single-trial classification of electroencephalographic signals. IEEE Transactions on Biomedical Engineering 51(6), 1052–1056 (2004)

Open Electrical Impedance Tomography: Computer Simulation and System Realization

Wei He, Bing Li[*], Chuanhong He, Haijun Luo, and Zheng Xu

State Key Laboratory of Power Transmission Equipment & System Security and New Technology, Chongqing University, Chongqing 400030, China
libing213@163.com

Abstract. Electrical impedance tomography (EIT) is a non-invasive technique used to image the electrical conductivity and permittivity within a body from measurements taken on body's surface. But the low spatial resolution of imaging, complicated operation and the asymmetrical placement of electrodes make the EIT cannot achieve the requirements of clinical application. In this paper, we proposed a novel idea of open electrical impedance tomography (OEIT) and a new type of electrode to provide more valuable distribution information of electrical impedance for local subsurface biological tissues than closed EIT. The OEIT system is constructed. The electromagnetic mathematical models of OEIT are presented and a method to solving the boundary value problem of half infinite calculating region is proposed for the computer simulations. Furthermore, we have realized the OEIT system, and completed the physics and the clinical experiments. Experiment results show that the OEIT system can obtain a better resolution and positioning accuracy, and is more suitable for clinical application.

Keywords: EIT, Open EIT, computer simulation, electrode array.

1 Introduction

Electrical impedance tomography (EIT) is particularly well-suited to applications where its portability, rapid acquisition speed and sensitivity give it a practical advantage over other medical monitoring or imaging technologies [1-2]. EIT technology has been researched for three decades, however it is almost at the laboratory stage all over the world, with only a few reports of clinical applications: TS2000 system of Israel [3], Angelplan-EIS1000 system of the Fourth Military Medical University [4], MEIK® system of Russia [5] and the MK 3.5 EIT system of The University of Sheffield [6] for breast cancer and pulmonary function diagnosis.

The main problem of EIT is that the resolution is worse than the mainstream medical imaging technologies like Computer Tomography (CT) and type-B ultrasonic. At present, the main pattern of EIT is using the electrodes surrounding the entire imaging object to measure the distribution of object's conductivity, which we call the Closed

---

[*] Corresponding author.

K. Li et al. (Eds.): LSMS / ICSEE 2010, Part I, CCIS 97, pp. 163–170, 2010.

Electrical Impedance Tomography (CEIT) [7]. Because CEIT simulation is based on the next two ideal assumptions: the simple circle imaging area and the symmetrical distribution of electrodes on the area boundary, the theoretical analysis and computer simulation are very precise. But in practical situations, the measuring objects have complex geometries, and the placement of electrodes cannot be symmetrical, the actual imaging results are hard to achieve the performance of the simulations. In order to improve EIT technology, we have proposed a novel EIT measuring pattern: the electrodes are only distributed in one side of the object, and the imaging area is considered to be a local area underneath the electrodes, which we call Open Electrical Impedance Tomography (OEIT). OEIT has the following advantages:

1) Electrode precise placement: because OEIT measures the local region of imaging object, all electrodes can be combined together, as a set of electrode array, to avoid the changes of electrode position when operating.

2) Easy operation: because the electrodes of OEIT are fixed as a set array, when operating, people only need to place it on the object at one time, avoiding the repeating operation of pasting electrodes in CEIT.

In this article, we study the principle of OEIT in computer simulation, which includes the forward problem and the inverse problem. In forward problem, we have considered the half infinite calculating region, and established the appropriate boundary value problem. And in inverse problem, we have utilized the most popular reconstruction algorithm: Newton's One-Step Error Reconstruction (NOSER) to image reconstruction [8]. Furthermore, we have realized the OEIT system, and completed the physics and the clinical experiments. In clinical experiments, we measured and imaged several breast cancer cases, and gotten some acceptable results.

2   Computer Simulation of OEIT

2.1   Calculated Region of OEIT

In CEIT, the imaging area is the cross section surrounded by the electrodes (shown in Fig.1a). So its calculated region is limited in a closed circle. But, the electrodes of OEIT are only placed on a part side of the object surface, so its calculating region is a half infinite area. However, the currents mainly distribute near the electrodes. The farther the point is, the smaller the current density is. So a virtual boundary is defined, on which the normal component of currents is zero, shown in Fig.1b.

2.2   The Boundary Value Problem

In EIT, the frequency of currents is low, so the boundary value problem is a quasi-static current field problem. We use the voltage potential φ as the variable of the controlling equation, which satisfies the Laplace equation [9]:

$$\nabla \cdot [\sigma(x, y) \cdot \nabla \varphi(x, y)] = 0 \qquad (x, y) \in \Omega \tag{1}$$

Where σ(x,y) is the conductivity distribution of the calculated region. While, in OEIT, the boundary is divided by two parts: the incomplete true boundary from which currents inject and outflow, and the virtual boundary on which the normal component of currents is assumed as zero.

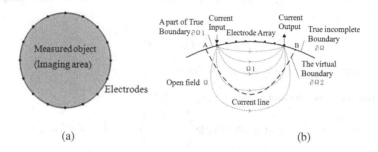

(a)   (b)

Fig. 1. Closed EIT and open EIT field model

$$\varphi(x, y) = f(x, y) \qquad (x, y) \in \partial\Omega1 \tag{2}$$

$$\sigma(x, y)\frac{\partial\varphi(x, y)}{\partial v} = j(x, y) \quad (x, y) \in \partial\Omega1 \tag{3}$$

$$\sigma(x, y)\frac{\partial\varphi(x, y)}{\partial v} = 0 \qquad (x, y) \in \partial\Omega2 \tag{4}$$

where f is potential of true boundary, j is current density injected into field Ω, v is outward normal unit vector of boundary.

Obviously, the boundary value problem of OEIT exists errors. The larger $\partial\Omega1$ and $\partial\Omega2$ are, the better the approximation closed to the actual situation, the larger the computational quantity is. So the key of simulation is to choose a suitable boundary.

2.3 The Inverse Problem

The OEIT's boundary value problem is solved by finite element method (FEM). The NOSER was developed for circular electrode geometries. It begins with a guess of an initial distribution of conductivities σ or resistivity ρ, where ρ_n is the reciprocal of σ_n, and n is the total number of mesh elements. Current patterns are applied to the 8 electrodes, and potentials are measured on each electrode V^k. Also, the potential on the electrodes $U^k(\rho)$ with the guess ρ can be computed. Our goal here is to minimize the sum of the squares of the differences between two potentials:

$$E(\rho) = \sum_{k=1}^{L-1}\left\|V^k - U^k(\rho)\right\|^2 = \sum_{k=1}^{L-1}\sum_{l=1}^{L}\left\|V_l^k - U_l^k(\rho)\right\|^2 \tag{5}$$

The minimization of $E(\rho)$ is equivalent to finding the zero point of the derivative of $E(\rho)$ as

$$F_n(\rho) = \frac{\partial E(\rho)}{\rho_n} = -2\sum_{k=1}^{L-1}\sum_{l=1}^{L}\left\|V_l^k - U_l^k(\rho)\right\|^2 \frac{\partial U_l^k(\rho)}{\partial \rho_n} \tag{6}$$

We then solve $F_n(\rho) = 0$. Using Newton's method, the solution can be obtained iteratively. The equation is

$$\rho_{new} = \rho_{old} - J_F^{-1}(\rho_{old})F(\rho_{old}) \tag{7}$$

The procedure can be summarized as follows.

1. Choose the best constant resistivity $\rho_{old} = c(1,1,\ldots 1)$.

2. Compute $\dfrac{\partial U_l^k(\rho_{old})}{\partial \rho_n}$, and approximate $\left\langle T^s, \dfrac{\partial U^k}{\partial \rho_n}\right\rangle \approx \dfrac{1}{\rho_n^2}\int_{M_n}\nabla u^k \bullet \nabla u^s$, where T^s is a

   set of orthogonal current patterns and M_n is the nth mesh element.

3. Approximate $J_F(\rho) = A_{n,m} + \gamma A_{n,m}\delta_{n,m}$, where $A_{n,m} = 2\sum_{k=1}^{L-1}\sum_{l=1}^{L}\dfrac{\partial U_l^k}{\partial \rho_n}\dfrac{\partial U_l^k}{\partial \rho_m}$, and $\delta_{n,m}$

   is a delta function, which equals 1 if n = m and zero otherwise. Gamma, γ is the regularization parameter, which should be chosen to be as small as possible.
4. Compute ρ_{new} by Newton's method and display the result on the chosen mesh.

2.4 Simulation Results

The calculated region is a 16cm×8cm rectangle, and the electrode array is placed on the upper surface. As shown in Fig.2a, the coordinate origin is fixed on the center of electrode array. To improve the forward problem accuracy and reduce the calculating quantity, we refine a 10cm×5cm area near the electrodes. The region conductivity is 1S/m. The target is a 5mm×5mm square object. There are two types of simulation: single target and double targets. In single target case, the conductivity is 0.2S/m. In double targets case, the conductivity of left one is 0.2S/m and right one is 5S/m.

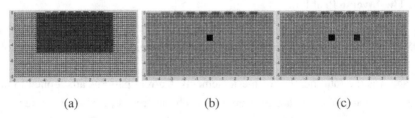

(a)   (b)   (c)

Fig. 2. Field model of OEIT simulation and the imaging target

Since the region is a symmetric rectangle in the horizontal direction, in the single target case, we only place it on the different positions of the left part of the region: two horizontal positions, -1cm and 0cm, and three vertical positions, -3cm, -2cm, -1cm. In the double targets case, the vertical position is the same as those of the single target case; the horizontal distances between two targets are 4cm and 1cm. The targets positions and the simulated results are shown in Fig.3. In single target simulation, the results indicate an obvious trend: the more the target is close to the surface or the center of electrode array, the better the image quality is. The simulation results in double targets are the same as that of the single target. The red area is the image of the target with larger conductivity.

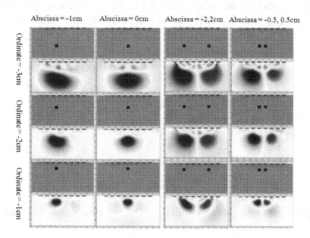

Fig. 3. Single target and double targets reconstruction result

3 System Realization

3.1 System Construction

The OEIT system construction, shown in the left of Fig.4, consists of the measuring probe and the image reconstruction computer, which are connected by USB cable. The electrode array is on the front of the measuring probe. It can inject current into human body and measure the voltage of the body surface. And the measuring circuit is integrated in the measuring probe, which transforms the analogy voltage to the digital data, and sends it to the computer. Then, the computer executes the image reconstruction program and displays the imaging results. A kind of composite electrode is used as shown in the right of Fig.4. The measuring electrode is bedded in the center of the stimulating electrode. This construction can reduce the negative impact of the contact impedance between skin and electrode.

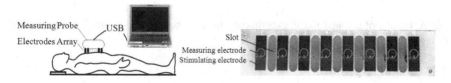

Fig. 4. OEIT system and Electrode Array

3.2 Physical Experiments

The electrochemical effect of agar is similar with biological tissue. So we use agar as the imaging object in physical experiments. The measuring probe is shown in Fig. 5a, which can be having a well contact with the imaging object. Fig.5b is the imaging object. The background is the white agar with the conductivity of 300 ms/m. The targets bedded in the background are two blocks of agar, the red one with the conductivity of 2s/m and the blue one with 20ms/m.

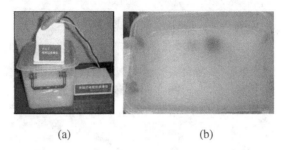

(a)   (b)

Fig. 5. The experiment system setup of OEIT (a: The experiment system; b: the agar model)

Fig.6a shows the reconstruction results of the low conductivity agar, while Fig.6b shows the high conductivity agar imaging. The three images in both figures are the results of target with 1cm, 2cm and 4cm distant from the electrode. The shade of the color presents the relative magnitude of the conductivity. The results show that the more the target is close to the surface, the better the image quality is. These results indicate that the OEIT system can be tested in the clinical experiment.

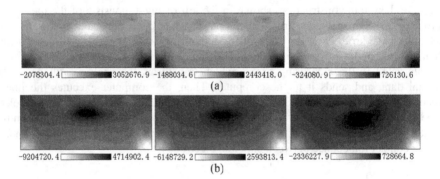

Fig. 6. Results of the agar model experiments. (a) Low conductivity target in agar model; (b) high conductivity target in agar model.

3.3 Clinical Experiment

In clinical experiments, we use the OEIT system to measure and imaging several breast cancer cases in Chongqing Cancer Hospital. Here is a typical case. The patient lay on the bed. The measuring probe was contacted with the breast. The imaging result is shown in Fig.7. We compared it with the results from X-ray mammography and ultrasound. In Fig.7a, the X-ray mammography finds out the tumor in the black circle area. In Fig.7c, the tumor is a 0.8×1.4cm hypoechoic area. In Fig.7b we can find a region with lower conductivity, which is considered to be the tumor by OEIT.

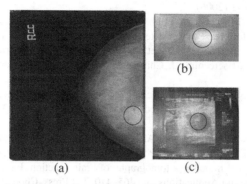

Fig. 7. Comparison of imaging results of breast carcinoma patient with three imaging methods (a: the result of X-ray mammography; b: the result of OEIT; c: the result of type-B ultrasonic)

4   Conclusions

This article proposes a novel EIT technology: open electrical impedance tomography, to break the limitation of CEIT. We have studied the principle of OEIT, and established its simulating model. The simulation has solved the boundary value problem of a half infinite region and achieved a desirable result. Furthermore, we realized the portable OEIT system and completed the physical and preliminary clinical experiments. PCB electrodes which could reduce contact impedance caused by the contact impedance between skin and electrode are used to instead of ECG electrode. According to the imaging results, the OEIT has acceptable performance, compared with X-ray mammography and ultrasound technologies.

Although OEIT has its own advantages, it is inevitable that a reduction in electrode number will lead to poorer image properties and an increased uncertainty in quantity estimations. Restricting electrodes to half the electrode plane further reduces overall image quality, increases spatial variance and therefore the variability in quantity estimates of an anomaly at an arbitrary location. In future work, we plan to take more clinical experiments and determine the effects on sensitivity, images and quantity estimates of artifacts such as breathing and movement.

Acknowledgments. This work was supported by The National Natural Science Foundation of China (50877082), Natural Science Foundation Project of CQ CSTC (CSTC2009BB5204), The project of scientists and engineers serving enterprise of the Ministry of Science and Technology of China (2009GJF10025), National "111" Project of China (Grant No: B08036).

References

1. Brown, B.H.: Electrical impedance tomography. J. Journal of Medical Engineering & Technology 27(3), 97–108 (2003)
2. Bayford, H.: Bioimpedance tomography (electrical impedance tomography). J. Annu. Rev. Biomed. Eng. 8, 63–91 (2006)
3. Michel, A., Orah, L.M., Dov, M., et al.: The T-SCAN^{TM} technology: electrical impedance as a diagnostic tool for breast cancer detection. J. Physiol Meas. 22(1), 1–8 (2001)
4. ZhenYu, J., XueTao, S., FuSheng, Y., et al.: Hardware design of electrical impedance scanning system for breast screening. J. Chinese Journal of Scientific Instrument 29(6), 1171–1175 (2008)
5. Jan, C., Nicola, D.: 3D EIT – MEIK in clinical application: observations and preliminary results. In: Magjarevic, R., Nagel, J.H. (eds.) IFMBE Proceedings, pp. 3906–3910. Springer, Berlin (2007)
6. Wilson, A.J., Milnes, P., Waterworth, A.R., et al.: Mk3.5: a modular, multi-frequency successor to the Mk3a EIS/EIT system. J. Physiol Meas. 22(1), 49–54 (2001)
7. Holder, D.S.: Electrical impedance tomography of brain function. In: Int. Cof. Electromagnetic Field Problems and Applications, pp. 465–470. TSI Press, Chongqing (2008)
8. Guanxin, X.: Theoretic and Applied Research of Electrical Impedance Tomography. Chongqing University, Chongqing (2004)
9. Wei, H., Ciyong, L., et al.: The principle of Electrical Impedance Tomography. Chinese Science Press, Beijing (2009)

Digital Watermarking Algorithm Based on Image Fusion

Fan Zhang[1,2], Dongfang Shang[2], and Xinhong Zhang[3]

[1] Institute of Image Processing and Pattern Recognition,
Henan University,
Kaifeng 475001, China
zhangfan@vip.sohu.com
[2] College of Computer and Information Engineering, Henan University,
Kaifeng 475001, China
[3] Computing Center, Henan University,
Kaifeng 475001, China
zxh@henu.edu.cn

Abstract. The process of watermarking can be viewed as a process of image fusion from the original image and the watermark image. Assuming that the watermarked image is the steady-state, Kalman filter is used as an optimal estimation algorithm in the process of image fusion. The math model is built according to the watermark image and the original image. Then the state equation and the corresponding measurement equation are built. The optimal estimation is received in the case of minimum estimation error variance. Experimental results show that the proposed algorithm has a good performance both in robustness and invisibility.

Keywords: watermarking, image fusion, Kalman filtering, mutual information.

1 Introduction

Digital watermarking techniques play an ever increasing role in the evolution information security. These roles include information integrity, security, privacy, ownership and other critical matters. Attention to these techniques has been increasing because of the ubiquity of digital records and the huge connectivity of individuals via large instantaneous networks[1]. Digital watermarking technology is an effective supplemental solution to the concern of digital media copyright protection. It embeds some marking information directly into the digital carrier (including multimedia, documents, software and so on), but it is not easily noticed by human perception (such as visual or auditory system) [2]. With the information hidden in carriers, it can confirm the contents, and helps the creator and the purchaser to transfer secret information and determine whether the carrier multimedia had been tampered [3].

K. Li et al. (Eds.): LSMS / ICSEE 2010, Part I, CCIS 97, pp. 171–176, 2009.
© Springer-Verlag Berlin Heidelberg 2009

2   Watermarking Based on Image Fusion

According to the principle of image processing and signal processing, digital watermarking can be viewed as a process that embedding a weak signal (watermark) to a strong signal (original image). Human Visual System (HVS) subjects to certain restrictions, such as the contrast threshold. The contrast threshold of HVS is affected by space, time and frequency characteristics. If the signal superposition is less than the contrast threshold of HVS, HVS will not able to feel the existence of the watermark signal. So it is absolutely feasible for the original image to embed some information without changing the visual effects.

Assume that S is the original image, F is the watermarked image (The watermarked image is also named as stego image) and W is the image of watermark, the process of watermarking can be expressed by the next formula,

$$F = S + f(S, W). \tag{1}$$

From the view point of image fusion, F could also be regarded as the image which integrated from the original image S and watermark image W, so the digital watermarking could be viewed as the process of image fusion. So, we can embed the watermark into the original image by using the technical of image fusion, such as Kalman filtering[4,5,6].

Kalman filtering has been a common method in information fusion, but it is rarely used in image watermarking. The target of this paper is to embed the watermark image W into the original image S, or to fuse the two images according to the image fusion technical. The simplest image fusion method is weighted average. Assume that the two images which need to fuse are W and S, and the fused image is F. The image fusion process of weighted average can be denoted as: $F = \omega_1 S + \omega_2 W$, where $\omega_1 + \omega_2 = 1$. ω_1 and ω_2 need to evaluate according to simulation experiments in the image fusion. For example, if the two images are very similar, the fused image can just be the average of the two images, then the ω_1 and ω_2 are 0.5 and 0.5 respectively. If the two images are very different, the fused image will choose the prominent one, then the ω_1 and ω_2 are 0 and 1 respectively.

3   Digital Watermarking Algorithm

Assume that S is the original image, its size is $m \times n$, and F is the optimal watermarked image, its size is $m \times n$. We assume the watermark image is W, and its size is also $m \times n$. Our work is to integrate the watermark into the original image. The target is to make the difference smallest between the embedded watermark image and the target image F. We expect achieve the optimal point between the robustness and the invisibility.

Let

$$X = \begin{bmatrix} F \\ S \end{bmatrix}, \tag{2}$$

where, $X \in \mathbb{R}^{2m \times n}$ is the state of the system to be estimated, $\mathbb{R}$ denotes the real domain. The system state remained unchanged in the observation process. We viewed the original image and the watermark image as the system state which are observed by two sensors at different time, then the system state equation and observation equation are as follows,

$$X(i+1) = X(i); i = 1, \tag{3}$$

$$Z(i) = H(i) X(i) + V(i); i = 1, 2, \tag{4}$$

where, $X(i) \in \mathbb{R}^{2m \times n}$ is the target state. $Z(i) \in \mathbb{R}^{m \times n}$ is the measurement of target state from i-th sensor. $H(i) \in \mathbb{R}^{m \times 2m}$ is the measurement matrix. $V(i) \in \mathbb{R}^{m \times n}$ is independent and zero-mean Gaussian white noise, and $R(i)$ is the variance of each element, it strictly satisfies,

$$R(i) = E[V_{kl}^2(i)]; k = 1, 2, \cdots, m; l = 1, 2, \cdots, n. \tag{5}$$

where $V_{kl}(i)$ is the element of $V(i)$ with k-th line and l-th row.

According to the Kalman filtering system (3) and (4), we have,

$$P(i+1|i) = P(i|i), \tag{6}$$

where $P(i+1|i)$ is the covariance matrix of estimate error in i time.

$$\hat{X}(i+1|i+1) =$$
$$\hat{X}(i+1|i) + P(i+1|i)H^T(i+1)[H(i+1)P(i|i)H^T(i+1)$$
$$+ R(i+1)]^{-1}[Z(i+1) - H(i+1)\hat{X}(i|i)], \tag{7}$$

$$P(i+1|i+1) =$$
$$P(i+1|i) - P(i+1|i)H^T(i+1)[H(i+1)P(i|i)H^T(i+1)$$
$$+ R(i+1)]^{-1}H(i+1)P(i+1|i). \tag{8}$$

Let

$$K(i+1) =$$
$$P(i+1|i)H^T(i+1)[H(i+1)P(i|i)H^T(i+1) + R(i+1)]^{-1}, \tag{9}$$

where $K(i+1)$ is Kalman gain matrix.

Then

$$\hat{X}(i+1|i+1) = \hat{X}(i|i) + K(i+1)[Z(i+1) - H(i+1)\hat{X}(i|i)], \tag{10}$$

$$P(i+1|i+1) = P(i|i) - K(i+1)H(i+1)P(i|i), \tag{11}$$

$$K(i+1) =$$
$$P(i|i)H^T(i+1)[H(i+1)P(i|i)H^T(i+1) + R(i+1)]^{-1}. \tag{12}$$

Let

$$\hat{F} = \hat{X}(2|2)^{(1:m)\times(1:n)}, \tag{13}$$

where $(1:m) \times (1:n)$ denotes that $\hat{F}$ is a part of matrix $\hat{X}(2|2)$ from the first column to the m-th column and from the first row to the n-th row. So when we embed watermark image into the original image, $\hat{F}$ is the optimal estimation of $F(2)$ in the case of the minimum estimation error variance. The selection of the parameters and the extracting method of the watermark are determined by the observation sequence of original and watermark image.

Assume that observed value $Z(1)$ is the original image and $Z(2)$ is the watermark image. The measurement matrix of (4) is

$$H(1) = \begin{bmatrix} O & I_{m\times m} \end{bmatrix}; \quad H(2) = \begin{bmatrix} I_{m\times m} & -I_{m\times m} \end{bmatrix}. \tag{14}$$

$I_{m\times m}$ is a unit matrix which the size is $m \times m$. The initial estimating measurement is

$$\hat{X}(1|1) = \begin{bmatrix} Z(1) \\ Z(1) \end{bmatrix}. \tag{15}$$

The covariance matrix of estimate error is

$$P(1|1) = \begin{bmatrix} R(1) & \\ & R(1) \end{bmatrix}. \tag{16}$$

The variance of noise is: $R(1)=0.01, R(2)=0.02$.

According to (10) and (12), we can obtain an estimate about the watermark,

$$\hat{X}(2|2) = [1 - K(2)H(2)]\hat{X}(1|1) + K(2)Z(2). \tag{17}$$

Because $K(2)$ is a matrix which the size

is $2m \times m$, it is irreversible, so watermark couldn't be extracted according to the calculation from (17). In (17), let

$$\begin{bmatrix} \hat{X}' \\ \hat{X}'' \end{bmatrix} = \hat{X}(2|2) - [1 - K(2)H(2)]\hat{X}(1|1);$$

$$\begin{bmatrix} K' \\ K'' \end{bmatrix} = K(2), \tag{18}$$

where $\hat{X}'$ and $\hat{X}''$ are the matrix which the size is $m \times n$, K' and K'' are the matrix which the size is $m \times m$. So (17) can also be rewritten as

$$\begin{bmatrix} \hat{X}' \\ \hat{X}'' \end{bmatrix} = \begin{bmatrix} K' \\ K'' \end{bmatrix} Z(2). \tag{19}$$

Then we have

$$\hat{X}' = K'Z\,(2)\,. \tag{20}$$

The watermark $\hat{W}$ can be extracted as follows,

$$W = K'^{-1}\hat{X}' \quad . \tag{21}$$

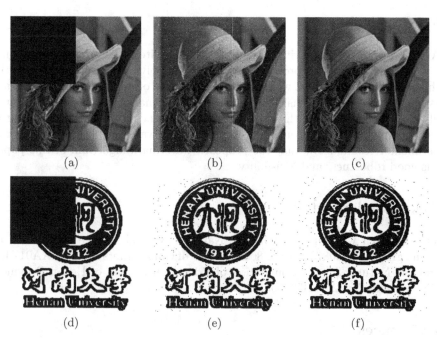

Fig. 1. The experimental result images under the different attacks. (a)(d) The cropping 1/4 of the experimental result image. (b)(e) Affected with pepper and salt noise. (c)(f) Affected with Gaussian noise.

4   Experiments

In this section, the images Lena, Peppers are used to demonstrate the effectiveness and the robustness of the proposed technique. And we evaluate the experimental results by the Mutual Information and the Cross Entropy. MATLAB 7.04 is used to simulate the watermarking process. The original image is 256 grayscale image with 256×256 pixels and the watermark image is a binary image with 256×256 pixels.

Fig. 1a shows the stego image after cropping 25% and Fig. 1d shows the corresponding extracted watermark. Fig. 1b shows the experimental result of stego image polluted by salt-and-pepper noise with the noise density 0.01. And Fig. 1e shows its corresponding extracted watermark image. Fig. 1c shows the experimental result of watermarked image polluted by Gaussian noise with the mean 0 and the variance 1, and Fig. 1f shows its corresponding extracted watermark image.

It's difficult to accurately compare the experimental results in different algorithms, because the experimental methods and parameters are different in different literatures, especially the bits of watermark and the strength of embedding are different. We can only compare our experimental results with other schemes approximately.

5 Conclusion

A digital watermark algorithm based on Kalman filtering is proposed. The purpose of this study is to propose a novel idea to solve watermark problems. In this watermark algorithm, we transform the process of watermarking to the state estimate process. Watermark is embedded in original image based on Kalman filtering. The novelty of this study is that the principle of Kalman filtering is used completely and thoroughly to design a novel watermarking algorithm. Experimental results show that the proposed watermark algorithm of this paper has good robustness and invisibility.

Acknowledgment

This research was supported by the National Natural Science Foundation of China grants No. 60873039, the 863 Program of China grants No. 2007AA01Z478 and the Natural Science Foundation of Education Bureau of Henan Province, China grants No. 2010B520003.

References

1. Maeno K, Sun Q, Chang S, Suto M. New semi-fragile image authentication watermarking techniques using random bias and nonuniform quantization. IEEE Transactions on Multimedia, 8(1): 32–45(2006).
2. Potdar V, Han S, Chang E. A survey of digital image watermarking techniques. in the Proceedings of the 3rd IEEE International Conference on Industrial Informatics, 709–716(2005).
3. Wang Sha, Zheng Dong, Zhao Jiying, Tam W. Speranza Filippo. An image quality evaluation method based on digital watermarking. IEEE Transactions on Circuits and Systems for Video Technology, 17(1): 98–105(2007).
4. Kalman R, Bucy R. New Results in Linear Filtering and Prediction Theory. Journal of Basic Engineering Transactions of the ASME, 83: 193–196(1961).
5. Khan A, Moura J. Distributing the Kalman filter for large-scale systems. IEEE Transactions on Signal Processing.56(10): 4919–4935(2008).
6. Toprak A, Guler I. Angiograph image restoration with the use of rule base fuzzy 2D Kalman filter. Expert Systems with Applications. 35(4): 1752–1761(2008).

The Dynamics of Quorum Sensing Mediated by Small RNAs in Vibrio Harveyi

Jianwei Shen[1,2] and Hongxian Zhou[1]

[1] Institute of Applied Mathematics, Xuchang University, Xuchang 461000, China
[2] Institute of System Biology, Shanaghai University, Shanghai 20044, China
jwshen8@yahoo.com.cn

Abstract. Quorum sensing (QS)is a important process of communication, we study a mechanism induced QS by coexist of small RNA and signal molecular (AI) in this paper. We construct a mathematical model to investigate phenomenon and find that there are periodic oscillation when the time delay and hill coefficient exceed a critical value. The periodic oscillation produces the change of concentration and induces QS. In addition, we also find the this network is robust against noise.

Keywords: Quorum sensing, Genetic network, Oscillation, Small RNA, Bifurcation, Negative feedback loop.

1 Introduction

The survival of bacteria relies mainly on regulatory networks which detect and integrate multiple environmental input and respond with appropriate behavioral output. These regulatory networks are involved in 'quorum sensing'(QS),which is a mechanism of chemical communication that enable bacteria to track population density by secreting and detecting extracellular signaling molecules called autoinducers (AIs)[5-7]. QS can regulate critical bacterial process by monitor the concentration of AIs, and alter the expression of large sets of genes to carry out task. Quantitative modeling of QS pathway can be useful to understand interaction between different genes.

In 1970, Nealson et al., had studied the model of the bioluminescent marine bacterium *Vibrio harveyi* for QS-based regulation, experimental studies indicate that there are multiple autoinducers and correponding sensor proteins which can regulate the expression of bioluminescent. The phosphorylation of regulator protein LuxO can be controled by the interaction between AIs and sensors. At negligible concentrations of AIs,i.e. at low cell density(LCD), these sensors act as kinases that transfer phosphate through LuxU to LuxO [9-11].LuxO P activates the expression of genes encoding small RNAs which in turn post-transcriptionally repress the QS master regulatory protein LuxR. At the high cell density (HCD), AIs accumulate and bind to their cognate sensors and the sensors act as phosphatases, reversing the phosphate flow through the QS circuit. This results in dephosphorylation and inactivation of LuxO,so that the expression of genes encoding the small RNAs is terminated.

K. Li et al. (Eds.): LSMS / ICSEE 2010, Part I, CCIS 97, pp. 177–184, 2009.
© Springer-Verlag Berlin Heidelberg 2009

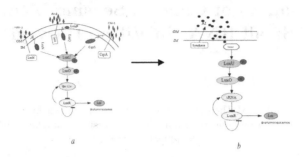

Fig. 1. Schematic diagram showing the gene regulation mediated by sRNA with a delayed negative feedback loop

Recent experiments[1,2] indicate that there is feedback loop between small RNAs and the QS master regulatory protein LuxR, LuxR directly activates transcription of genes encoding the small RNAs and inhibits the expression of genes encoding its self. The feedback loop involving in small RNA is essential for optimizing the dynamics of transition between individual and group behaviors. Due to the complexity of this regulatory network, we would desire to develop a quantitative framework for explaining the mechanism how to optimizing the dynamics of transition between individual and group behaviors. The corresponding quantitative model can then be used to make testable predictions for future experiments as well as to further analyze existing experimental data. The rest of the paper is organized as follows. In section 2, we give an overview of the QS network in *V.harveyi* and develop a mathematical model of QS network mediated by small RNAs and define some dimensionless parameters. In section 3, connect our model to experimental results. In section 4,

2 Formulation of the Model

The quorum-sensing network in *V.harveyi* is shown in figure 1. There are three pathways to control LuxU. The principal components of the pathway are three sensor (LuxN, LuxPQ, CqsS) and the corresponding autoinducer synthase(LuxM, LuxS, CqsA) which can produce the three autoinducers :H-AI1, AI-2 and CAI-1,respectively. The binding of the single autoinducer to a sensor is highly specific. the interaction between autoinducers (AI) and sensor proteins can determine the overall phospherylation state of LuxO through the phosphorelay mechanism. the phosphorylation state of LuxO can determine the activation state of corresponding small RNAs (sRNAs),Qrr2-4. These small RNAs are small noncoding RNAs, 18-24 nt in length, that are predicted to regulate the expression od approximately one-third of all human genes. This regulation occurs posttranscriptionally through small RNA binding to mRNA targets leading to target degradation or inhibition of translation. Experimental data [1-2] that phosphorylated LuxO can activate the expression of small RNAs, and the master regulatory protein *LuxR* is the target protein of sRNA (i.e Qrr2-4) and can activate the the qrr (genes of sRNAs) promoters, in the same time, $LuxR$ can autorepress the expression of itself (see figure 1). To understand the mechanism of the quorum sensing mediated by small RNAs, we consruct a mathematical model abstracted from the complex network (figure 1 and figure 2). In figure 2, we simplify the process of QS and abstract the model from figure 1, the three sensors (LuxN,LuxPQ,CqsS),Synthases(LuxM, LuxS,CqsA) autoinducers (HAI-1,AI-2,CAI-1) and sRNAs (Qrr2-4) can be viewed as one component.

In this QS pathway, the sensor protein can be modeled as two-state systems [14,15], i.e, kinase sate and phosphatase state. In the kinase state, the sensor (S_k) can autophosphorylate and then transfer the phosphate group to the downstream protein LuxU, whereas in the phosphatase state, the phosphate flow is reversed. From the experiments we can know that the sensors are in the kinase state at LCD, whereas at HCD, the sensors are primarily in phosphatase state.So we can consider a network wherein the free sensor corresponds to the kinase state(S_k), whereas sensors binded by autoinducers are in phosphatase state(S_p). According to the chemical reaction, we can obtain the chemical reaction equation as follows.

$$M \xrightarrow{k_1} AI, \ nAI \underset{k_{-2}}{\overset{k_2}{\rightleftharpoons}} A_n, \ A_n + S_k \underset{k_{-3}}{\overset{k_3}{\rightleftharpoons}} S_p, \ S_k + U \xrightarrow{k_4} S_k + U_p, \ U_p + S_p \xrightarrow{k_5} S_p + U, \ U_p +$$

$$O \underset{k_{-6}}{\overset{k_6}{\rightleftharpoons}} U + O_p$$ In the above reaction, M represents synthase which produces the corresponding autoinducer, AI represents autoinducer,S_k represents sensor corresponding to the kinase state, S_p represents the sensor corresponding to the phosphatase state. U represents $LuxU$, U_p represents phosphorylated $LuxU$. O represents $LuxO$ and O_p represents phosphorylated $LuxO$.

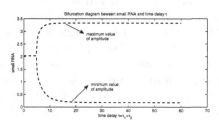

Fig. 2. Bifurcation diagram with total time delay τ as a parameter

By using the Mass Action Theory and Michaelis-Mentens Kinetics, we can obtain the mathematical model as follows

$$\frac{d[LuxM]}{dt} = a_1 - k_1[LuxM]$$

$$\frac{d[AI]}{dt} = k_1[LuxM] - d_1[AI]$$

$$\frac{d[LuxU_p]}{dt} = k_4[S_k](U_0 - [LuxU_p]) - k_5[S_p][LuxU_p] - d_3[LuxU_p]$$

$$\frac{[sRNA]}{dt} = \frac{1 + k_7[protein]^m(t - \tau_1) + \delta_1[LuxO_p]}{1 + [protein]^m(t - \tau_1) + \delta_2[LuxO_p]} - r[sRNA][mRNA] - d_4[sRNA]$$

$$\frac{d[mRNA]}{dt} = \frac{k_8}{1 + [protein]^m(t - \tau_1)} - r[sRNA][mRNA] - d_5[mRNA]$$

$$\frac{d[protein]}{dt} = k_9[mRNA](t - \tau_2) - d_6[protein]$$

$$\frac{d[Lux]}{dt} = \frac{k_{10}[protein]^m(t - \tau_1)}{1 + [protein]^m(t - \tau_1)} - d_7[Lux]$$

$$(1)$$

where $[LuxO_p] = \frac{[LuxU_p]O_0}{k[U_0]+(1-k)[LuxU_p]}, k = \frac{k_{-5}}{k_5}, [O_0] = [LuxO] + [LuxO_p], U_0 = [LuxU] + [LuxU_p], S_0 = [S_k] + [S_p], S_p = \frac{[AI]^n[S_0]}{\kappa^n + [AI]^n}, S_k = \frac{\kappa^n[S_0]}{\kappa^n + [AI]^n}, \kappa = (\frac{k_{-2}k_{-3}}{k_2 k_3})^{\frac{1}{n}},$ $[mRNA]$ and $[protein]$ represent the concentration of the master regulator HaR mRNA and protein,respectively, $[Lux]$ represents the concentration of luciferase which is required for bioluminescence.

3   Oscillation Dynamics of Quorum Sensing Model

3.1   The Dynamics with Time Delay

Delay often appears in the process of gene regulation, and affects the dynamics of gene network. From the theory of bifurcation, we can obtain a critical value τ_0 of time delay $\tau = \tau_1 + \tau_2$ and the other condition which the system is stable. As a example, assume that $a1 = 1, k1 = 0.2$, $d_1 = 0.5$ $k_5 = 0.3, s_0 = 2, u_0 = 2, d_2 = 0.6, k_4 = 0.3, \kappa = 0.8, n = 2$, $k_6 = 1, r = 1, d_3 = 0.5, \delta_1 = 1, \delta_2 = 1.1$, $d_3 = 0.5, d_4 = 0.3$, $k_8 = 2, d_5 = 0.2, k_9 = 1, d_6 = 0.1, k_{10} = 1, d_7 = 0.8, O_0 = 2, k = 0.6$. as the time delay$\tau$ changes, the dynamics of the system is changed. There is a critical value τ_0, when $\tau > \tau_0$,the steady state becomes unstable and the sustained oscillations occur, in the meantime, we also find the robustness of amplitudes against variation in delays.So we can control the dynamics of the system by tuning the time delay. In the system, delay τ_1 and τ_2 represent transportation or diffusion process from nucleus to cytoplasm of mRNA and from cytoplasm to nucleus of protein, respectively. From the analysis, we can know that τ_1 and τ_2 affect the dynamical behaviors in the form of $\tau_1 + \tau_2$ due to the cyclic structure of system. In addition, with the increase of τ, the period will also increase(see Fig.3).

3.2   The Dynamics with Hill Coefficient m

In the process of gene regulation, protein as TF(Transcription Factor) is often multi-polymerization, the degrees of polymerization often affect the dynamical behaviors. as a example, we make use of parameters in section 3.1. In the steady state, the concentration of small RNA, mRNA, HapR protein and Lux is changed with the variance of Hill coefficient m. With the increase of m, a bifurcation occurs at a critical value$m_0 \simeq 7.5$ (see Fig.4) and there will be periodic oscillation.

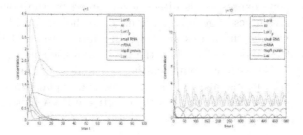

Fig. 3. Time history diagram with different time delay

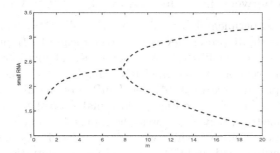

Fig. 4. Bifurcation induced by the hill coefficient m

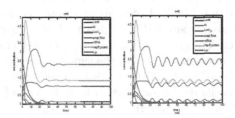

Fig. 5. Time history diagram with the hill coefficient m

In this case, the amplitude of oscillation will be also changed, and the maximum of amplitude also increases. At the same time, the period will also increase with the increases of Hill coefficients m due to increase of time of hill coefficient m (See Fig.5).

3.3   The Dynamics with Others

In this paper, we also study the change of concentration with the base pairing rate r. Obviously, the concentration of small RNA, mRNA, HapR protein and Lux will decrease with the increase of r. So the small RNA plays an important role in repressing the expression of target gene. So we can control the expression of target mRNA by regulating the base pairing rates.

In addition, gene expression is often accompanied by the noise, the noises often affect the dynamics of gene regulation, so we should consider these noise. In this paper, we take the noises as Gaussian white noise, and find that this network can not be almost affected by the noises and the small RNA can filter high frequency noise without compromising the ability to rapidly respond to large changes in input signals. In this case, we can explain this phenomenon due to a large pool of sRNAs shortens the effective mRNA lifetime and buffers against target mRNA fluctuations [16].

4   Mechanism of Quorum Sensing

4.1   Relation between hapR mRNA and Small RNA Levels

More and more experimental data show that small RNA, hapR mRNA and master regulator HapR protein levels vary reciprocally[1], HapR protein activates the expression of small RNAs. As τ increases and goes beyond the critical value τ_0, a limit cycle occurs, so small RNA, hapR mRNA and HapR protein levels vary reciprocally(see Fig.6(b)).The LapR proteins directly activate the transcription of small RNA (see Fig.6) which makes up the deficiency at HCD mode, at the same time, they also repress the expression of mRNA which lead to the downregulation of mRNA which verify the results of experimental data [1].

At first, the system (1) is in LCD mode and the concentration of small RNAs increases which induce the decrease of the concentration of mRNA, in the meanwhile, the HapR protein actives the expression of small RNA and represses the expression of mRNA (see Fig.6). In the steady state, as time delay τ exceeds τ_0, there is attractor among small RNA, mRNA, HapR protein and Lux, which attracts to a limit cycle.Thus it creates a periodic cycle from LCD to HCD mode. In this mechanism of quorum sensing, as the signal molecules (AI) first increase and then decrease, the sRNA-LuxR feedback loop activates the transition from LCD to HCD and accelerates the transition from HCD to LCD. According to analysis of Section 3, we can know that this feedback loop can filter the noise, so this is very good mechanism and it may be the result of natural selection.

4.2   Transition from Coexistence of $LuxO_p$ and Protein

During the transition from the low to high cell density mode, AI levels first increase and the cell switch from $LuxO_p$ dominated LCD mode to HCD mode dominated by HapR protein. During this switch, $LuxO_p$ and HapR protein could transiently coexist, which allow HapR protein feedback-activates small RNAs(see Fig.7(b)). When the time delayτ of transportation or diffusion from nucleus to cytoplasm of mRNA and from cytoplasm to nucleus of protein and the Hill coefficient m in Eq.(1) exceed the critical value, the concentration of small RNA, mRNA, protein,Lux changes periodically, which make up the deficiency of increase of AI. and these delay the transition from LCD mode to HCD mode. When AIs begin decreasing, these periodical oscillation accelerates the transition from HCD mode to LCD mode. These periodical oscillation result from the Hopf bifurcation induced by time delayτ and Hill coefficient m.

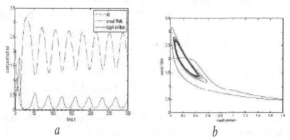

Fig. 6. hapR mRNA and small RNA levels vary reciprocally and hapR mRNA levels influence small RNA levels

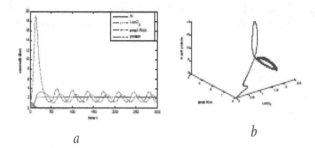

Fig. 7. Protein feedback on the small RNA in the presence of protein and $LuxO_p$

5   Conclusions

The bioluminescent marine bacterium often uses a cell-cell communication process called quorum sensing (QS)to co-ordinate behaviors in response to changes in population density. In this paper, we investigate the dynamics of a gene network which can induce the QS phenomenon. In this gene network, Changes in the concentration of Small RNA, HapR protein will lead to the change in the concentration of bioluminescent protein Lux by time delay and the hill coefficient. From the theoretical analysis and simulation, we can know that the the coexistence of AI and small RNA produce this kind of QS. In this mechanism, the system is robust against the noise.

Acknowledgments. This work is supported by NSF of China (10802043, 10832006) and Program for Science & Technology Innovation Talents in Universities of Henan Province (2009HASTIT033).

References

1. Tu, K.C., Waters, C.M., et al.: Molecular Microbiology 70(4), 896–907 (2008)
2. Svenningsen, S.L., Waters, C.M., Bassler, B.L.: Gene & Developent 22, 226–238 (2008)

3. Newman, M.E.J.: SIAM Review 45, 167–256 (2003)
4. Banik, S.K., Fenley, A.T., Kulkarni, R.V.: Physical Biology 6, 046008 (2009)
5. Miller, M.B., Bassler, B.L.: Annual Review of Microbiology 55, 165–199 (2001)
6. Waters, C.M., Bassler, B.L.: Annual Review of Cell and Developmental Biology 21, 319–346 (2005)
7. Bassler, B.L., Losick, R.: Cell 125, 237–246 (2006)
8. Nealson, K.H., Platt, T., Hastings, J.W.: Journal of Bacteriology 104, 313–322 (1970)
9. Freeman, J.A., Bassler, B.L.: Journal of Bacteriology 181, 899–906 (1999)
10. Freeman, J.A., Bassler, B.L.: Molecular Microbiology 31, 665–677 (1999)
11. Lilley, B.N., Bassler, B.L.: Molecular Microbiology 36, 940–954 (2000)
12. Gartel, A.L., Kandel, E.S.: Seminars in Cancer Biology 18, 103–110 (2008)
13. Mendell, J.T.: Cell 133, 217–222 (2008)
14. Neiditch, M.B., Federle, M.J., Pompeani, A.J., et al.: Cell 126, 1095–1108 (2006)
15. Swem, L.R., Swem, D.L., Wingreen, N.S., Bassler, B.L.: Cell 134, 461–473 (2008)
16. Mehta, P., Gojal, S., Wingreen, N.S.: Molecular Systems Biology 4, 221 (2008)

An Algorithm for Reconstruction of Surface from Parallel Contours and Its Section Contour Extraction in any Cutting Plane

Chun Gong, Can Tang, Yanhua Cheng, Sheng Cheng, and Jianwei Zhang

Kunshan Industrial Technology Research Institute, Jiangsu, P.R. China
{gongchun,tangcan,chengyanhua,chengsheng,
zhangjianwei}@ksitri.com

Abstract. To obtain reconstruction of surfaces from a given contours stack, this paper presents a new algorithm based on MC-algorithm, which solves the problem that 3D surface model cannot be built because the first/last contour has no previous/next contour information, or there are only isolated contours. Meanwhile, the algorithm is applied to extracting section contour of the 3D model in any cutting plane. The algorithm does not need to do as following: traverse and rebuild each triangle of the 3D model, obtain the points of intersection with the cutting plane, and link all points of intersection to get the contour in cutting plane. The algorithm divides the cutting plane into rectangular grid and gets isoline in each marching rectangle directly, without considering the problem of connections of all points of intersection. The 2D sectional contour can be obtained directly after finishing calculating isoline in the rectangular grid.

Keywords: cross-sectional contours, MC-algorithm, Three-dimensional surface reconstruction.

1 Introduction

To produce 3D model from planar contours is a very important problem in visualization, which is widely used in medical image visualization and terrain data visualization. At present there are a lot of algorithms for reconstruction of single-contour has a lot, including based on the global optimum search strategy [1, 2] and based on local calculations [3, 4] and so on. A.B.Ekoule[5] et al solved the reconstruction of non-convex planar single contour. After that, single contour reconstruction has been relatively mature. But more reconstruction arises from more contours in different slice. The problem is much more difficult than that of reconstruction of single contours. The contour stitching approach to surface reconstruction attempts to generate a surface by connecting the vertices of adjacent contours [6]. These methods need to address the branching, tiling and correspondence problems. Generally, these problems are complicated.

Another method to deal with the problem is volumetric method. W. M. Jones [7] et al introduced a distance fields for each contour, then the problem for reconstruction of surface becomes a problem of getting the zero isosurface of the volume. By using

K. Li et al. (Eds.): LSMS / ICSEE 2010, Part I, CCIS 97, pp. 185–193, 2010.

well-known MC-algorithm, the isosurface in every cube can be extracted from the volume. To obtain reconstruction of surface between two slices, the method required that there are 2D contours in each slice. Therefore, the method could not solve such case that there are contours in one slice, but no contours on last or next slice. Distance fields on the slice that has no contours can not be calculated. Furthermore, taking the Euclidean distance that those points to the contours on current slice as the points' distance fields is not accurate enough and can not be well used for reconstruction of surface, for the distance is not exact distance to 3D model reconstructed from planar contours. Therefore, current method for calculating distance fields should be improved.

Given these defects, a new method is presented to calculate the more precise distance fields. On one hand the method can deal with the problem of how to get distance fields that slice have no contours but next or last slice has; on the other hand, the distance fields calculated will be more precise than current method [7]. Then we use the MC-algorithm [8] to produce the final reconstructed surface.

Since the volumetric surface reconstruction method can be used to build 3D model from planar contours by using MC-algorithm, then it can be used to obtain 2D section contours in any cutting plane of the 3D model by using 2D MC-algorithm.

2   Volumetric Surface Reconstruction

2.1   The Distance Fields Computation

The input of the algorithm is finite planar contours in the slices $z_0 \ldots z_n$. Let Ω be the 3D model, the Ω_i be the distance fields of z_i.

$$\Omega_i = \{ f(x,y) | (x,y,z_i) \in \Omega \} . \tag{1}$$

Generally the exact distance field computation is very time-consuming, then the first step of the algorithm is to determine whether it's necessary to calculate the distance field of the point (x,y,z). Field function can be used to solve the problem; the field function is defined as

$$f(x,y) = \begin{cases} -1, (x,y) & is & outside & all & contours \\ 0, (x,y) & is & on & a & contour \\ 1, (x,y) & is & inside & a & contour \end{cases} . \tag{2}$$

If field function of every voxel's apex is -1 or 1, and then there are no isosurface in the voxel, it is not necessary to calculate distance field of all the voxel's apexes. Otherwise, Distance field should be calculated. The distance field is defined by W. M. Jones et al as

$$f(x,y) = \begin{cases} -dist(x,y), (x,y) & is & outside & all & contours \\ 0, (x,y) & is & on & a & contour \\ dist(x,y), (x,y) & is & inside & a & contour \end{cases} . \tag{3}$$

Here dist(x, y) is the minimal distance from (x, y) to the contours in current slice plane. However, the minimal distance from (x, y) to the 3D model results from the distance that the point to contours in next or last slice sometimes. Obviously, the

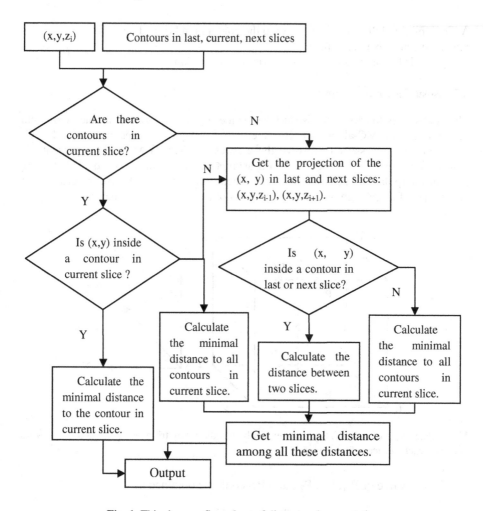

Fig. 1. This shows a flow chart of distance of computation

definition of distance field is not accurate. A new method to define the distance field is given as

$$dist(x, y) = \begin{cases} \min(CurrD_{\min}, \sqrt{AdD_{\min}^2 + SliceD^2}) \cdots & case \quad A \\ \\ CurrD_{\min} & \cdots\cdots \quad case \quad B \end{cases} \quad (4)$$

Case A: if(x, y) is outside all contours.
Case B: if(x, y) is inside a contour.
$CurrD_{\min}$: distance from (x, y) to the contours (case A) or a contour which the (x, y) is inside a contours in current slice (case B).

AdD$_{min}$: minimal distance from (x, y) to the contours in last and next slice.
SliceD: distance between slices.

Figure1 is the process of distance of field computation.

2.2 Isosurface Extraction

The next step is to extract the isosurface for the scalar value zero from the voxel data set in terms of MC-algorithm. But there are ambiguous cases in standard MC-algorithm, it's necessary to resolve these cases. Nielson G M et al [9] present the criteria for connecting vertices base on asymptote in one plane of voxel. By using trilinear interpolation, the value in a voxel according to values of voxel's vertices can be obtained. Figure2 demonstrates the process of calculating the value of P.

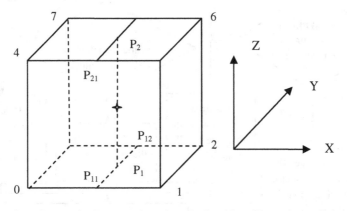

Fig. 2. The value of any point in a voxel can be calculated by trilinear interpolation. $a_0 \sim a_7$ are values of each vertices of the voxel.

Firstly, the value of P_{11}, P_{12}, P_{21}, and P_{22} could be obtained as

$$P_{11}= a_0 + (a_1 - a_0)\,x$$
$$P_{21}= a_4 + (a_5 - a_4)\,x$$
$$P_{12}= a_3 + (a_2 - a_3)\,x \tag{5}$$
$$P_{22}= a_7 + (a_6 - a_7)\,x$$

Then P_1, P_2 can be gotten as

$$P_1 = P_{11} + (P_{12} - P_{11})y = a_0 + (a_1 - a_0)x + (a_3 - a_0)y + (a_0 + a_2 - a_1 - a_3)xy$$
$$P_2 = P_{21} + (P_{22} - P_{21})y = a_4 + (a_5 - a_4)x + (a_7 - a_4)y + (a_4 + a_6 - a_5 - a_7)xy \tag{6}$$

Finally, the value of P can be calculated as

$$P = a_0 + (a_1 - a_0)x + (a_3 - a_0)y + (a_4 - a_0)z + (a_0 + a_2 - a_1 - a_3)xy + (a_0 + a_5 - a_1 - a_4)xz$$
$$+ (a_0 + a_7 - a_3 - a_4)yz + (a_1 + a_3 + a_4 + a_6 - a_0 - a_2 - a_5 - a_7)xyz \tag{7}$$

A0~A7 can replace the coefficient of (7), the value of P could be described as (8).

$$P = A_0+A_1x+A_2y+A_3z+A_4xy+A_5xz+A_6yz+A_7xyz \tag{8}$$

From (8) asymptotes in all grid planes of the voxel can be obtained. For example, in the plane $Z = Z0$, the asymptotes of isovalue hyperbolas are:

$$X = - (A2 + A5Z0)/ (A4+A7Z0)$$
$$Y = - (A1 + A6Z0)/ (A4+A7Z0)$$

By this way, ambiguous cases can be eliminated. After inputting every vertices' distance field of a voxel, isosurface can be extracted.

2.3 2D Marching Cubes

Sometimes, doctors want to carefully observe the sectional contours of some biological tissue in any cutting plane, such as coronal slice, sagittal slice. However, these slice images can not be provided by imaging devices directly. Therefore, some algorithms have been presented to deal with the problem [10-11].

But all these methods are related to 3D surface reconstruction, while the process of 3D surface reconstruction is generally time-consuming. In fact, 2D Marching Cubes can be used to solve the problem rapidly. It's the method of extracting the isoline in every mesh rectangle. There are 16 (=2^4) ways which the isoline may intersect the rectangle. The symmetry of rectangle makes the number down from 16 cases to 6 as Fig 3.

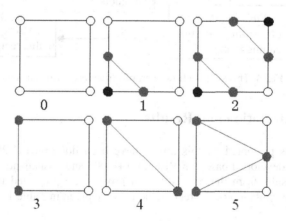

Fig. 3. Different cases of isoline in rectangles. There are ambiguous cases for Case 2.

Even if the cases are less than MC-algorithm, there are ambiguous cases (case 2 figure3) too. It's necessary to use trilinear interpolation to deal with these cases, and another four points come from corresponding points in one of the adjacent cutting plane. The algorithm of extraction of 2D sectional contour of 3D model in any cutting plane can be described as the flow chart (Fig 4).

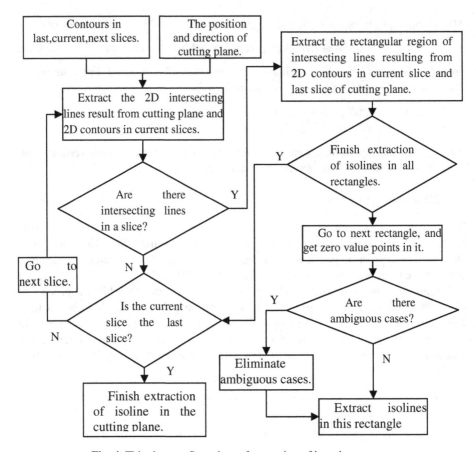

Fig. 4. This shows a flow chart of extraction of isovalue contour

3   Experimentation and Results

All algorithms presented by this article have been done on PC 2GB RAM. These programs are developed based on Visual C++ 2005 and Visualization Toolkit version 5.0.4. The results from the reconstruction process are presented in Figures 5. The datasets for computation of the reconstructions are given in Table 1.

Table 1. Dataset of contours in every slice and computation times

Distance ratio of X:Y:Z per pixel	# of slices	Datasets in every contour in two adjacent slices
1:1:7.58	1	(90,74) (96,66) (102,67) (103,72) (103,80) (98,81) (92,78)
	2	(90,74) (96,66) (102,67) (103,72) (103,80) (98,81) (92,78)

Table 2. Computation time of two algorithms

Algorithm Type	Construction of Table1 Computation times (milli-seconds)
Presented by W. M. Jones et al	313
This article	359

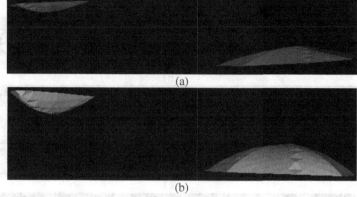

(a)

(b)

Fig. 5. Algorithm for surface of reconstruction by W. M. Jones et al (a) and this article (b)

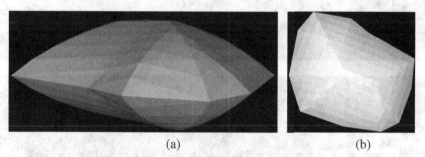

(a)   (b)

Fig. 6. Surface reconstruction results from isolated single contour in one slice that has no contours in adjacent slices. See from front (a) viewport and top viewport (b).

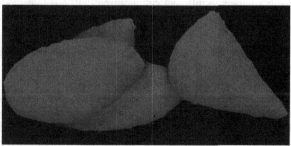

Fig. 7. 3D reconstructed surface model based on MC-algorithm, distance field calculated as flow chart Fig 1

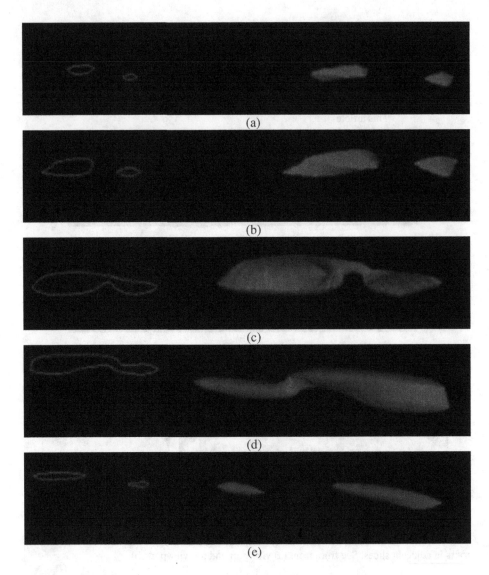

Fig. 8. 2D isovalue contours in different cutting plane (left) and remainder 3D model (right) covered by the cutting planes. (a) ~ (c) from front side, (d), (e) from inverse side

The resulting surfaces (see Fig. 5) demonstrate that this contour-based surface reconstruction approach produces more smooth models than that produced by the algorithm by W. M. Jones et al. In addition, 3D model' shape won't change when the distance of contours in adjacent slice is much more than size of these contours.

A 3D model are presented in Fig. 7, which results from the reconstruction of three contours, and two contours in one slice and one in another adjacent slice. There are 512×512 pixels in every slice. Coronal cutting planes divide every slice into 100 parts. The results of extracted isovalue contours of the 3D model in any cutting plane by algorithm described in Figure 4 are presented in Figures 8(a) ~ (e).

4 Conclusion and Future Work

This article presents a volumetric approach to reconstruct a complete 3D surface from a sparse set of parallel contours. This algorithm especially aims at solving the problem that 3D surface model cannot be built because the first/last contour with no previous/next contour information. It presents a new method for calculating 2D distance fields, which makes distant fields in those slice have no contour can be obtained. The algorithm for surface reconstruction of adjacent two slices has no restriction so that even isolated single contour can be used to build complete 3D surface. Meanwhile, combined with 2D MC algorithm, the algorithm can be used to extract the cross-sectional contours in any cutting plane just after finishing the process of computation of zero line in every grid rectangles in the cutting plane. Therefore, it is unnecessary for calculating and linking all points of intersection with 3D surface model to get cross-sectional contours. In the future, the main focus is about how to improve the efficiency of the computation of 2D distance fields and get smoother reconstructed surface at the same time.

References

1. Fuchs, H., Kedem, Z.M., Uselton, S.P.: Optimal surfaces reconstruction from planar contours. Communication of the ACM 20(10), 693–702 (1977)
2. Keppel, E.: Approximating complex surfaces by triangulation of contour lines. IBM Journal of Research Development 19(1), 2–11 (1975)
3. Christiansen, H.N., Sederberg, T.W.: Conversion of complex contour line definitions into polygonal element mosaics. Computer Graphics 12(3), 187–192 (1978)
4. Ganapathy, S., Dennehy, T.G.: A new general triangulation method for planar contours. Computer Graphics 16(3), 69–75 (1982)
5. Ekoule, A.B., Peyrin, F.C., Odet, C.L.: A triangulation algorithm from arbitrary shaped multiple planar contours. ACM Transactions on Graphics 10(2), 182–199 (1991)
6. Meyers, D., Skinner, S.: Surfaces from Contours. ACM Transactions on Graphics 11(3), 228–258 (1992)
7. Mark, W., Jones, C.M.: A new approach to the construction of surfaces from contour data. Euro Graphics 13(3), 75–84 (1994)
8. Lorensen, W.E., Cline, H.E.: Marching Cubes: A High Resolution 3D Surface Construction Algorithm. Computer Graphics 21(4), 163–169 (1987)
9. Nielson, G.M., Hamann, B.: The Asymptotic Decider: Resolving the Ambiguity in Marching Cubes. In: IEEE Proceedings of Visualization, pp. 83–91 (1991)
10. Lotjonen, J., Reiss man, P.J., Magnin, I.E., et al.: A triangulation method of an arbitrary point set for biomagnetic problems. IEEE Transactions on Magnetics 34(4), 2228–2233 (1998)
11. Dolenc, A., Makela, L.: Slicing procedures for layered manufacturing techniques. Computer aided Design 26, 119–126 (1994)

Simulation Modeling of Network Intrusion Detection Based on Artificial Immune System

Yu Jing and Wang Feng

PLA Artillery Academy, Hefei, 230031, China

Abstract. There is much comparability between natural immune system and computer security, and the key point is how to distinguish self from other. Based on the principles and structures of artificial immune system, the simulation modeling of network intrusion detection was developed. The model consisted of many nodes that distributed across different locations. The nodes needed not be centralized controlled. The purpose of model was to distinguish between illegitimate behavior (non-self) and legitimate behavior (self). In case of finding abnormity, model could give an alarm to user.

Keywords: artificial immune system; instruction detection; computer security; detector.

1 Introduction

Intrusion detection system (IDS) is a dynamic security protection technology and is the second line of security defense behind firewall. Most of commercial IDS have realized

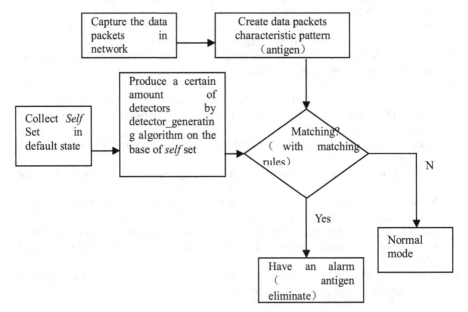

Fig. 1. Flow chart of intrusion detection system from immunology

K. Li et al. (Eds.): LSMS / ICSEE 2010, Part I, CCIS 97, pp. 194–199, 2010.
© Springer-Verlag Berlin Heidelberg 2010

the function of character detection by far. Dynamic abnormity detection should be study emphasis of intrusion detection.

The researches of IDS based on immunology have practical significance. It is likely to be the new method to break away from puzzledom currently [1, 4]. The paper applied immunology to IDS and analyzed the requirement of network system and the characteristics of natural immune system. The model monitored head data of traffic packets in a broadcast local area network. Using dynamic detectors to impersonate T-cells, a kind of IDS simulation model based on immunology was designed (see Fig.1).

2 The Simulation Modeling of Network Intrusion Detection Based on Immunology

2.1 Formalization Description of the Simulation Model

The definition was made in a given computer network environment:

(1). P: Set of all patterns in network stream. The environment of the model was defined as a universe set P, and P was partitioned into two subsets, S and N, which called self and non-self.

(2). S: Self, described the patterns which were harmless, acceptable, legal to network computer system [2].

(3). N: Non-self, described the patterns which were harmful (endangering the integrality, validity, secrecy of computer system), unacceptable, unlawful to network computer system. Respectively $S \cap N = \varnothing$, $S \cup N = P$.

(4). The problem faced by an intrusion detection system was: Given limited resources, classified a pattern p ($p \in P$) as normal, corresponding to self, or anomalous, corresponding to non-self.

(5). Mapping function λ

$\lambda : P \to P_r$. Any pattern p of universe set P was mapped to only binary machine figure. Universe set P, S (self) and N (non-self) were mapped to P_r, S_r, N_r by mapping function λ, and $S_r \cup N_r = P_r$, $S_r \cap N_r = \varnothing$, p which appeared in network stream stochastic was mapped to q. That was to say, λ (P) $= P_r$, λ (S) $= S_r$, λ (N) $= N_r$, λ (p) $= q$.

(6). Model definition of network intrusion detection based on immunology [2]: The distributed environment was defined by a finite set L of locations, and the number of locations was n. That was $|L| = n$. $\forall l \in L, M_l, D_l, f_l, h_l$ were detection node, detectors set, classification function and the matching rules of the local node l respectively, and that was described $M_l = (D_l, f_l, h_l)$.

①. Classification function of local node: the model was a distribute system, and the different detection nodes could communicate resource and detect cooperate with each other. The detection nodes worked on the same principle with immunity cell. So the classification function of local node was described: $\forall l \in L, M_l$

$$f_l(D_l, h_l, q) = \begin{cases} anomalous, & if \ \forall d \in D_l \ and \ h_l(q,d) = 1 \\ normal, & otherwise \end{cases}$$

②. Detectors set of local node: Detectors had memory which was the encounter intrusion similar to immunity cell once. D_l was the detectors set of local node l, and $D_l = Dm_l \cup Di_l \cup Du_l$. Dm_l, Di_l, Du_l were representative for memory detectors, mature detectors, immature detectors according to detection priority level. The detector amount was same of every node and detectors could adjust themselves to the detection results incessantly.

③. Classification function of system: P_l was a representative for the network link patterns which appeared in node l.

$$g(\{M_l\}, q) = \begin{cases} anomalous, & if \ \exists l \in L, \ q \in P_l \ and \ f_l(D_l, h_l, q) = anomalous \\ normal, & otherwise \end{cases}$$

Classification function of system described the detection ability of the system. According to the function, if one classification function of any local node judged a pattern q to be non-self, classification function of system would also judge a pattern q to be non-self. In other words, the model of network intrusion detection based on immunology was analogue that system of which detectors was $\bigcup_{j \in L} M_j$.

④. The matching rules of every node were same or different, and that was $h_i = h_j$ or $h_i \neq h_j$.

2.2 Function Design of the Simulation Modeling

2.2.1 Design of Distributed Detection Node
A distributed detection node was designed as follows (see fig.2):

(1). Capturing data package module: Destination hosts only captured a packet copy in order to not influence normal users in the net section.
(2). The filtration module: The task of this module was to filter the WWW and FTP service.
(3). Formation and creation antigen module: The antigen consisted of characteristic information of data package.
(4). Judgment and regular update detector module: The detectors impersonated T-cells, so the detectors set in model could adjust themselves to the network state and update automatically. In other words, detectors could self-learn and self-adjust.
(5). Alarm and communication module: The task of this module was to alarm and notice other detection nodes.
(6). Detectors producing and initialize module: This module was responsible for bringing into the initial detectors [3].
(7). Detectors training module [3, 4]: This module was responsible for training the initial detectors in a period of time, and it made the initial detectors spend bearing phase and appear a various state.

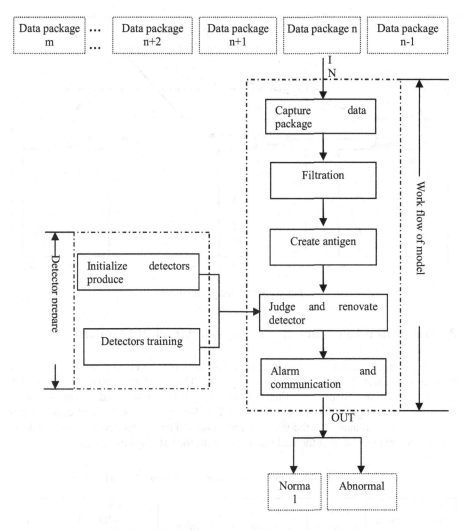

Fig. 2. A detection node design in modeling

2.2.2 Design of Whole Model
The whole simulation model was designed as follows (see fig.3):

(1). Main body of system was distributed, and every node was a distributed detection node structure that we designed in 2.2.1.
(2). Every node had many detectors (similar to the immune cells) of fixed quantity to check the data package and give the alarm. Every detector set of local node was not identical completely, so the non-self that could be detect was also not identical completely. But nodes could communicate each other [5, 6].
(3). Every node was self-governed, and the system needed not the central controller.
(4). The antigen was smaller and could represent as many as possible characteristic intrusions, and it was easy to realize.

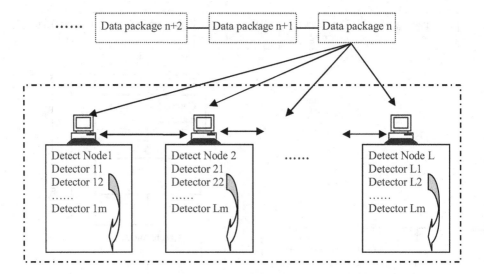

Fig. 3. Whole design of model

3   Simulation Experiments of Non-self Attacking Model

A binary string structure from the information of network data packages to describe the current secure status of the network was explored. The structure consisted of source IP, destination IP, source port, destination port, length of packet, and so on. We used Winsock technological to complete the model in Windows LAN. We had done several intrusion tests and analyzed the test data and results. The results demonstrated that the system model could detect the real intrusions with the little system cost.

Table 1. Simulation experimentations of non-self attack

Intrusion event	Port scanning	Finite port scanning	Single port scanning	SYN FLOOD attack	Back orifice
Real connection number	542	543	2345	745	121
Alert number	436	399	1785	605	113
Error alert number	52	75	74	39	5
Detection rate	71%	58%	73%	75%	89%

Analysis of experiments result:

(1). According to the data of the table1, detection rate was not always consistent with theory result. The reason was that the binary strings of self and non-self were not completely random but mostly definite density distribution. The detection rate would fluctuate due to distributing density.

(2). The detection rate of which character was not distinct (for example, finite port scanning) was relative low.
(3). If the alternation time of single port scanning was relative long, or if the ports were separate, it would be not propitious to system detection.

4 Conclusions

A simulation modeling of network intrusion detection based on immunology was described and analyzed. The significant work of this paper was as follows:

(1). A simulation modeling of network intrusion detection based on immunology was developed.
(2). The formalization description of the simulation modeling was explored.
(3). The function design of the simulation modeling was carried out, including design of distributed detect node and design of whole modeling.

The characteristic patterns were produced only from the head of data packet, so the content of packet will be the focal point that we next work. In order to simplify mathematical analysis and test, match length r was fixed [7], but it was probably more reasonable that r is various.

References

1. Forrest, S., Hofmeyr, S.A.: Computer Immunology (DRAFT). Communications of the ACM 40(10), 88–96 (1997)
2. Hofmeyr, S.A.: An Immunological Model of Distributed Detection and its Application to computer Security. In: Genetic and Evolutionary Conference (GECCO 1999), Orlando, Florida, A late_breaking Paper, July 25-27, pp. 149–158 (1999)
3. Kim, J., Bentley, P.J.: An Evaluation of Negative Selection in an Artificial Immune System for Network Intrusion Detection. In: Proceedings of the Genetic and Evolutionary Computation Conference (GECC), pp. 1330–1337 (2000)
4. Kim, J., Bentley, P.: The Artificial Immune Model for Network Intrusion Detection. In: 7th European Conference on Intelligent Techniques and Soft Computing, EUFIT 1999 (1999)
5. Somayaji, A., Hofmeyr, S., Forrest, S.: Principles of a computer immune system. In: Proceeding of New Security Paradigms Workshop 1997, Langdale,Cumbria, pp. 75–82 (1998)
6. Hofmeyr, S., Forrest, S.: Architecture for an Artificial Immune System. In: Evolutionary Computation, vol. 7(1), pp. 1289–1296. Morgan-Kaufmann, San Francisco (2000)
7. Dasgupta, D., Forrest, S.: Artificial Immune Systems in Industrial Applications. Accepted for Presentation at the International Conference on Intelligent Processing and Manufacturing Material (IPMM), Honolulu, HI, July 10-14 (1999)

Organic Acid Prediction in Biogas Plants Using UV/vis Spectroscopic Online-Measurements

Christian Wolf[1], Daniel Gaida[2], André Stuhlsatz[3], Seán McLoone[1], and Michael Bongards[2]

[1] National University of Ireland Maynooth, Department of Electronic Engineering,
Co. Kildare, Maynooth, Ireland
[2] Cologne University of Applied Sciences, Institute of Automation and Industrial IT,
Steinmüllerallee 1, 51643 Gummersbach, Germany
[3] Düsseldorf University of Applied Sciences, Institute for Information Technology,
Department of Mechanical and Process Engineering,
Josef-Gockeln-Str. 9, 40474 Düsseldorf, Germany
{christian.wolf,daniel.gaida,michael.bongards}@fh-koeln.de,
andre.stuhlsatz@fh-duesseldorf.de, sean.mcloone@eeng.nium.ie

Abstract. The concentration of organic acids in anaerobic digesters is one of the most critical parameters for monitoring and advanced control of anaerobic digestion processes, making a reliable online-measurement system absolutely necessary. This paper introduces a novel approach to obtaining these measurements indirectly and online using UV/vis spectroscopic probes, in conjunction with powerful pattern recognition methods. An UV/vis spectroscopic probe from S::CAN is used in combination with a custom-built dilution system to monitor the absorption of fully fermented sludge at a spectrum from 200nm to 750nm. Advanced pattern recognition methods, like LDA, Generalized Discriminant Analysis (GerDA) and SVM, are then used to map the measured absorption spectra to laboratory measurements of organic acid concentrations. The validation of the approach at a full-scale 1.3MW industrial biogas plant shows that more than 87% of the measured organic acid concentrations can be detected correctly.

Keywords: LDA, GerDA, SVM, classification, UV/vis spectroscopy, organic acids, online-measurement, anaerobic digestion.

1 Introduction

Recent developments in the area of online-measurement and advanced control systems show the importance of using powerful computational intelligence and data analysis methods to exploit the full potential of such systems [1]. The application of feature extraction and classification methods to predict organic acid concentrations in anaerobic digestion processes using UV/vis spectroscopic probes is an example of such a hybrid system.

In particular, monitoring and control of anaerobic digestion in biogas plants has proven to be extremely difficult due to a lack of robust and feasible online-measurement

K. Li et al. (Eds.): LSMS / ICSEE 2010, Part I, CCIS 97, pp. 200–206, 2010.
© Springer-Verlag Berlin Heidelberg 2010

systems and the high non-linearity of anaerobic digestion processes. Nevertheless, it becomes more important than ever to offer solutions for advanced process control of biogas plants, because efficient plant operation is the major issue when it comes to feasible long-term operation of such plants. The robust measurement of organic acid concentrations, being one of the key parameters for process stability, is required to develop and test innovative optimization and control strategies for anaerobic digestion processes. The availability of UV/vis spectroscopic probes offers a new approach to measuring organic acid concentrations indirectly and online. By employing powerful feature extraction and classification methods, organic acid concentrations can be predicted from the absorption spectra measurements taken from diluted fermentation sludge. In this paper we consider the well-known Linear Discriminant Analysis (LDA) and the Generalized Discriminant Analysis (GerDA), which is a novel and powerful extension of the classical LDA algorithm [2], to extract features automatically from the raw measurements. Based on that, linear classifiers are used to classify the extracted features into different concentration ranges. For comparison, the use of Support Vector Machines (SVM) on the raw measurements is investigated as well.

2 Materials and Methods

2.1 The Dataset

The spectrometric measurement device provides a characteristic absorption curve, called fingerprint, over $p \in \mathbf{N}$ wavelengths. The values are given in $[Au/m]$ and stored as a column vector, where the ith one is denoted by $\mathbf{x}_i \in X$, with the feature space $X \subseteq \mathbf{R}^p$. In total we have $N \in \mathbf{N}$ such fingerprints, i.e. $i = 1, \ldots, N$. Associated with each such vector $\mathbf{x}_i$ is the ith organic acid sample with unit $[g/l]$, denoted by $c_{a,i} \in \mathbf{R}$. To formulate the mapping from $\mathbf{x}_i$ to $c_{a,i}$ as a classification problem the measurements $c_{a,i}$ are clustered into $C = 5$ classes, which account for the whole range of given c_a's. The class $\vartheta \in \Theta$ to which the ith organic acid measurement belongs to is given by $\vartheta_i \in \Theta$, where $\Theta := \{1,2,3,4,5\}$ are the class labels as defined in table 1. The size of the data set obtained from the biogas plant, its separation into training and validation data sets and the distribution of the samples across classes are also illustrated in table 1.

Table 1. Definition of the class labels and the number of samples in each class ϑ for the complete (N_ϑ), training ($N_{T,\vartheta}$) and validation dataset, $N_{V,\vartheta}$

Class $\vartheta \in \Theta$	Organic acid concentration $c_a [g/l]$	N_ϑ	$N_{T,\vartheta}$	$N_{V,\vartheta}$
1	1.1, ..., 1.4	228	171	57
2	1.5, ..., 1.8	1528	1146	382
3	1.9, ..., 2.2	1880	1410	470
4	2.3, ..., 2.6	731	549	182
5	2.7, ..., 3.0	70	52	18

From an initial investigation with the data set containing the total spectrum using LDA it was determined that better results could be obtained by omitting the longer wavelengths, hence a new data set was created, such that the LDA results were optimized with respect to p. The resulting reduced feature space data set, is capped at 640nm leading to a $p = 176$ dimensional feature vector x_i. Experiments have shown, that the results obtained with this new data set, were also better using GerDA and SVM as the results gained with the original data set.

2.2 Linear Discriminant Analysis (LDA)

Linear Discriminant Analysis searches for a linear transformation $\mathbf{A} \in \mathbf{R}^{m \times p}$, $m \leq p$, such that the transformed data $\mathbf{Y} = \mathbf{A} \cdot \mathbf{X}$, $\mathbf{Y} := (\mathbf{y}_1, \ldots, \mathbf{y}_{N_T}) \in \mathbf{R}^{m \times N_T}$, can be linearly separated better than the original feature vectors $\mathbf{X} := (\mathbf{x}_1, \ldots, \mathbf{x}_{N_T}) \in \mathbf{R}^{p \times N_T}$. The linear transformation $\mathbf{A}$ is determined by solving an optimization problem maximizing the well-known Fisher discriminant criterion:

$$trace\{\mathbf{S}_T^{-1} \cdot \mathbf{S}_B\} \tag{1}$$

with the common total scatter-matrix $\mathbf{S}_T$ and between-class scatter-matrix $\mathbf{S}_B$ [3].

The LDA and a subsequent linear classifier are both implemented in MATLAB® [4]. An LDA transformation into a feature space of $m = C - 1 = 4$ dimensions led to the best subsequent linear classification results.

2.3 Generalized Discriminant Analysis (GerDA)

LDA is a popular preprocessing and visualization tool used in different pattern recognition applications. Unfortunately, LDA and a subsequent linear classification produce high error rates on many real world datasets, because a linear mapping $\mathbf{A}$ cannot transform arbitrarily distributed features into independently Gaussian distributed ones. A natural generalization of the classical LDA is to assume a function space F of nonlinear transformations $f : \mathbf{R}^p \to \mathbf{R}^m$ and to still rely on having intrinsic features $\mathbf{Y} := f(\mathbf{X})$ with same statistical properties as assumed for LDA features. The idea is that a sufficient large space F potentially contains a nonlinear feature extractor $f^* \in F$ increasing the discriminant criterion (1) compared to a linear extractor $\mathbf{A}$.

GerDA defines a large space F using a Deep Neural Network (DNN), and consequently the nonlinear feature extractor $f^* \in F$ is given by the DNN which is trained with measurements of the data space such that the objective function (1) is maximized. Unfortunately, training a DNN with standard methods, like back-propagation, is known to be challenging due to many local optima in the considered objective function. To efficiently train a large DNN with respect to (1), in [2], [5] a stochastic pre-optimization has been proposed based on greedily layer-wise trained Restricted Boltzmann Machines (RBM) [6].

After layer-wise pre-optimization all weights $\mathbf{W}$ and biases $\mathbf{b}$ of the GerDA-DNN are appropriately initialized. Nevertheless, pre-optimization is suboptimal in maximizing (1), thus a subsequent fine-tuning of the GerDA-DNN is performed using a modified back-propagation of the gradients of (1) with respect to the network parameters. In [2], [5] it is shown that stochastic pre-optimization and subsequent fine-tuning yields very good discriminative features and training time is substantially reduced compared to random initialization of large GerDA-DNNs.

For the extraction of intrinsic features from the raw measurements, we used GerDA with a p-250-50-25-m topology, i.e. a 5 layer DNN consisting of one input layer with p units, 3 hidden layers with 250, 50 respectively 25 units and one output layer with m units resulting in more than 265 million of free parameters. To avoid the effect of over-fitting of the training data, we terminated the fine-tuning after the pre-training stage using an early-stopping criterion dependent on the training error. The topology of GerDA as well as the early-stopping criterion was evaluated on the training data via 5-fold cross-validation. Additionally, the best intrinsic dimensionality $m \geq C - 1$ was cross-validated, too. The GerDA-framework is implemented in MATLAB®.

2.4 Support Vector Machines (SVM)

Support Vector Machines offer a computationally efficient method for multi-class classification problems by finding hyperplanes, which separate data sets into classes in a high dimensional feature space. For the classification problem under consideration a C-Support Vector Classification is used with soft margin optimization and a Radial Basis Function Kernel (RBF Kernel) [7] using the SVM implementation LIBSVM [8].

Due to the biased data set, SVMs initially performed poorly on class 5. This was addressed by using a weighted SVM with a weighting factor of 100 given to class 5 compared to 1 for all the other classes.

2.5 Practical Application

The online measurement of organic acid concentrations as key parameter is important to detect and solve problems quickly to guarantee stable and efficient plant operation. High organic acid concentrations decrease the pH level in the bioreactor and cause high stress on methane-producing bacteria, which are no longer able to process the available substrate. Such a change in environmental conditions may easily lead to a complete collapse of the anaerobic digestion process as shown in Figure 1.

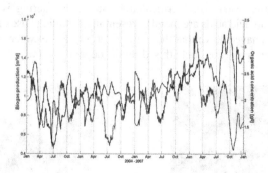

Fig. 1. Collapse of biogas production due to rising organic acid concentration at an industrial biogas plant

The state-of-the-art way to measure and monitor organic acid concentration unfortunately still is to perform laboratory analysis of the fermentation sludge and substrate feed on a regular basis, which makes fast process monitoring and control nearly impossible. Based on the current situation in the field of online-measurement systems for biogas plants, the need for new developments in sensor technology and engineering solutions in this area is huge. Existing technologies for online-measurements of organic acid concentrations in biogas plants, such as gas-phase chromatographs or automatic titrators are only available to a small group of biogas plant operating companies as they are expensive and high-maintenance products.

The new approach discussed in this paper is to use UV/vis spectroscopy, which uses ultraviolet light (200nm - 750nm) to determine the concentration of a certain substance in a liquid sample. The main problem for the application on biogas plants is the high concentration of organic acids in the substrate and also the relatively high concentration of solids. Thus, an automated sample preparation and dilution system has been developed that addresses these issues and which is installed on an industrial biogas plant near Gummersbach, Germany. With an electrical power of 1.3MW this plant uses biological municipal waste for fermentation.

Conducted laboratory tests with the S::CAN spectro::lyser show that organic acid concentrations can be detected by analyzing the absorption over several wavelengths as proven by Henning et al [9]. The results indicate that organic acid concentrations cannot be measured separately but as a composite parameter, which makes UV/vis spectroscopy well suited for organic acid measurement on biogas plants. Furthermore, the absorption beyond 300nm is very low, which leads to the conclusion that absorption characteristics at higher wavelengths do not have a high impact on the concentration of organic acids. However, wavelengths above 300nm are needed to take account of substrate coloring and the changing matrix of mains water in the measurement. Therefore, as discussed in Section 2.1, the appropriate cut-off point was estimated by selecting the feature vector length that yielded the best performance with LDA.

Online-measurement Apparatus
Due to the fact, that total solids (TS) concentration in the digester is up to 20%, a direct measurement of the absorption of the substrate at different wavelengths is nearly impossible, as the 1mm gap width of the UV/vis probe (S::CAN spectro::lyser) is easily soiled. For this reason, it is necessary to build up a special dilution system for the fermentation sludge. In this case, water from fermentation sludge dewatering is used for online-measurements, as organic acids are mainly present in the liquid phase of the sludge.

Laboratory tests have shown that the optimal ratio between water and sample is 1:80 to get a clear spectrum. To reach this dilution degree for an accurate measurement

Fig. 2. a. UV/Vis-probe with 1mm gap width; **b.** Layout of the measurement system

with the S::CAN spectro::lyser probe, the dilution unit is filled with four litres of water every 30 minutes (batch process). A flexible-tube pump is used to administer a defined amount of the fermentation press water (50ml). The following figures show the layout of the measurement and dilution system.

3   Classification Results

To validate and compare the classification performances of the different methods we decided to use an unbiased measure for validation, which is determined by the mean of the diagonal elements of the confusion matrices (see table 2) and therefore is independent of the number of samples N_ϑ. Nevertheless for the calibration of the methods the standard mean validation error is used.

As can be seen in table 3 GerDA achieves the best results, measured by the unbiased performance measure, followed closely by the SVM implementation.

Table 2. Confusion matrices

LDA		predicted					GerDA		predicted			
[%]	1	2	3	4	5		[%]	1	2	3	4	5
1	**68.4**	14.0	8.8	8.8	0.0		1	**98.3**	0.0	0.0	0.0	1.8
given 2	7.1	**64.9**	20.2	6.0	1.8		given 2	3.1	**91.6**	4.2	0.8	0.3
3	1.9	17.0	**71.1**	8.7	1.3		3	0.0	4.5	**88.7**	4.0	2.8
4	1.6	17.0	30.8	**42.3**	8.2		4	1.1	3.3	12.1	**68.7**	14.8
5	0.0	5.6	5.6	5.6	**83.3**		5	0.0	0.0	11.1	0.0	**88.9**

SVM		predicted			
[%]	1	2	3	4	5
1	**93.0**	5.3	1.7	0.0	0.0
given 2	2	**92.4**	5	0.6	0.0
3	0.2	6	**89.4**	3.2	1.2
4	2.7	4.4	6	**78.6**	8.3
5	0.0	5.5	5.6	11.1	**77.8**

Table 3. Results

Feature Extractor	Classifier	Validation error [%]
LDA	linear	34.0
GerDA	linear	12.8
none	SVM	13.7

4   Discussion and Conclusion

This paper demonstrates a new approach to successfully measure organic acid concentrations online using UV/vis spectrometric measurements, which offers new possibilities for advanced plant operation and control. The close monitoring of anaerobic digestion processes and the development of control strategies for optimal organic acid concentrations will substantially increase process efficiency and stability. However, results show that this online-measurement is far from trivial, such that advanced pattern recognition methods are needed to achieve good results. With the use of SVM and a novel method named Generalized Discriminant Analysis[1] (GerDA) results with a 13% error rate can be achieved, which is sufficiently accurate to be of value for the online-measurement of organic acids. GerDA yields the best error rate of the methods considered (12.8%) and further has many desirable properties. The GerDA-framework is self-contained and easy to use with learning performed in a semi-supervised manner. It can be used as a substantial dimension reduction step for a fast and simple classification of high-dimensional multi-class data.

It became obvious that the non-uniform distributed class sizes hamper the pattern recognition methods, especially the implementation and configuration of the SVM used here. To prevent this kind of problem it is important to use a performance measure, which is unbiased for validation and also for determination and optimization of method parameters. This is expected to lead to further improvement of GerDA and SVM respectively and therefore will further improve the organic acid online-measurement system.

References

1. Puñal, A., Palazzotto, L., Bouvier, J.C., Conte, T., Steyer, J.P.: Automatic control of VFA in anaerobic digestion using a fuzzy logic based approach. In: IWA VII Latin American Workshop and Symposium on Anaerobic Digestion, pp. 119–126 (2002)
2. Stuhlsatz, A., Lippel, J., Zielke, T.: Discriminative Feature Extraction with Deep Neural Networks. In: Proceedings of the 2010 International Joint Conference on Neural Networks (IJCNN), Barcelona, Spain, pp. 18–23 (2010)
3. Duda, R.O., Hart, E., Stork, D.G.: Pattern Classification. In: John W., Sons. (eds.) Inc. (2000)
4. Matlab®, The Mathworks, http://www.mathworks.com (last accessed 25.04.2010)
5. Stuhlsatz, A., Lippel, J., Zielke, T.: Feature Extraction for Simple Classification. In: Proceedings of the International Conference on Pattern Recognition (ICPR), Istanbul, Turkey, pp. 23–26 (2010)
6. Hinton, G., Osindero, S., Teh, Y.W.: A Fast Learning Algorithm for Deep Belief Nets. Neural Computation 18(7), 1527–1554 (2006)
7. Cristianini, N., Shawe-Taylor, J.: An Introduction to Support Vector Machines and other kernel-based learning methods. Cambridge University Press, Cambridge (2000)
8. Chang, C.C., Lin, C.J.: LIBSVM: a library for support vector machines (2001), http://www.csie.ntu.edu.tw/~cjlin/libsvm
9. Schmidt, H., Rehorek, A.: New Online-Measurement Methods for Biogas Plants. Diploma Thesis at the Cologne University of Applied Sciences. Cologne (2008)

---

[1] Available as MATLAB® code (Request at andre.stuhlsatz@fh-duesseldorf.de).

Power Quality Disturbances Events Recognition Based on S-Transform and Probabilistic Neural Network

Nantian Huang[1,2], Xiaosheng Liu[1], Dianguo Xu[1], and Jiajin Qi[3]

[1] Department of Electrical Engineering, Harbin Institute of Technology,
Harbin, China
huangnantian@126.com
[2] College of Information and Control Engineering, Jilin Institute of Chemical Technology,
Jilin, China
[3] Hangzhou Electric Power Bureau, State Grid Corporation of China, Hangzhou, China

Abstract. Power quality (PQ) events recognition is the most important research area of power quality control. A novel high performance classification system based on S-transform and probabilistic neural network is proposed in this paper. Firstly, S-transform processes the original PQ signals into a complex matrix named S-matrix. The time and frequency features of disturbances signal are extracted from the S-matrix. Then, the selected subset of features is used as the input vector of the classifier. Finally, the probabilistic neural network classifier is trained and tested by the simulated simples. The simulation results show the effectiveness of the new approach.

Keywords: power quality (PQ), power quality disturbances, S-transform, probabilistic neural network.

1 Introduction

With the use of more modern electronic equipments and automation, power quality (PQ) and its automatic monitoring and analysis has become an important challenging issue for power engineers [1]. PQ disturbances can be classified into two main categories, "variation type" and "event type". They are identified during monitoring by exceedance of a defined threshold and characterized by a set of appropriate parameters for each event [2]. Topical disturbances mainly include voltage sags, swells, interruption, flicker, harmonic and transients. The disturbances events with short duration and numerous types are hard to recognize and control.

Power quality disturbances events recognition is the foundation of PQ disturbances analysis and control. The traditional identification process includes two steps, such as feature extraction and pattern recognition. Time-frequency analysis methods could appear the detail features from time domain and frequency domain. Compare to STFT [3], DWT and WPT [4], the S-transform [5] is an extension of wavelet transform, and it is based on a moving and scalable localizing Gaussian window. It has characteristics superior to either of the time-frequency transforms in PQ analysis area. The classification system based on S-transform and different structures of neural networks [6-8] have already used for PQ events identification with good effect.

K. Li et al. (Eds.): LSMS / ICSEE 2010, Part I, CCIS 97, pp. 207–212, 2010.

This article proposes a new method based on S-transform and Probabilistic Neural Network (PNN) for power quality disturbances recognition. Firstly, the 8 types of disturbance signals including 2 types of complex disturbances are established by mathematical model and the simulated signals are obtained for training and verification the classifier. Then the appropriate features extracted from the result of S-transform are used to constitute input vector of PNN to train the classifier. At last, the trained classifier based on PNN is used for power quality disturbances recognition.

1.1 The S-Transform

S-transform is an extension of continuous wavelet transform (CWT). It provides frequency-dependent resolution with a direct relationship of the Fourier spectrum. The advantages of the S-transform are due to the fact that the modulating sinusoids are fixed with respect to the time axis, whereas the localizing scalable Gaussian window dilates and translates [5].

The S-transform of a time series $h(t)$ is defined as

$$S(\tau, f) = \int_{-\infty}^{\infty} h(t)g(\tau - t, f)e^{-i2\pi ft}dt \tag{1}$$

where f is the frequency, τ and t are both time.

The Gaussian modulation function $g(\tau, f)$ is given as

$$g(\tau, f) = \frac{|f|}{\sqrt{2\pi}} e^{-(t^2/2\sigma^2)} \tag{2}$$

and $\sigma = \dfrac{1}{|f|}$. The final expression is defined as

$$S(\tau, f) = \int_{-\infty}^{\infty} h(t) \left\{ \frac{|f|}{\sqrt{2\pi}} e^{-((\tau - t)^2 f^2/2)} e^{-i2\pi ft} \right\} dt \tag{3}$$

S-transform of a discrete time series $h(t)$ is derived by letting $\tau \to jT$ and $f \to n/NT$ is like

$$S(jT, \frac{n}{NT}) = \sum_{m=0}^{N-1} H[\frac{m+n}{NT}]G(m,n)e^{i2\pi mj/N} \tag{4}$$

where $G(m, n) = e^{-2\pi^2 m^2 \alpha^2/n^2}$, $j, m, n = 0, 1, ..., N-1$.

1.2 Probabilistic Neural Network (PNN)

The probabilistic neural network (PNN) proposed by Specht [9] is an outgrowth of the Bayesian classifier with Parzen window. In the signal-classification area, the training examples are classified according to their distribution values of probabilistic density function (pdf) [10, 11]. A simple pdf is defined as follows:

$$f_k(\mathbf{x}) = \frac{1}{N_k} \sum_{j=1}^{N_k} \exp(-\frac{\| \mathbf{x} - \mathbf{x}_{kj} \|}{2\sigma^2}) \tag{5}$$

Modifying and applying (5) to the output vector $\mathbf{H}$ of the hidden layer in the PNN is as follows:

$$H_h = \exp(\frac{-\sum_i (\mathbf{x}_i - W_{ih}^{xh})^2}{2\sigma^2}) \tag{6}$$

$$net_j = \frac{1}{N_j} \sum_h W_{hj}^{hy} H_h \ \text{and} \ N_j = \sum_h W_{hj}^{hy} \tag{7}$$

$$net_j = \max_k(net_k), \text{then} \ y_j = 1, \text{else} \ y_j = 0. \tag{8}$$

Where i is the number of input layers; h is the number of hidden layers; j is the number of output layers; k is the number of training examples; N is the number of classifications (clusters); σ is the smoothing parameter (standard deviation); $\mathbf{x}$ is the input vector; W_{ih}^{xh} is the connection weight between the input layer $\mathbf{x}$ and the hidden layer H ; W_{hj}^{hy} is the connection weight between the hidden layer H and the output layer Y . $\| \mathbf{x} - \mathbf{x}_{kj} \|$ is the Euclidean distance between the vectors $\mathbf{x}$ and $\mathbf{x}_{kj}$, where $\| \mathbf{x} - \mathbf{x}_{kj} \| = \sum_i (\mathbf{x} - \mathbf{x}_{kj})^2$.

Each pattern unit contributes to its associated category unit a signal equal to the probability the test point was generated by a Gaussian centered on the associated training point. The sum of these local estimates (computed at the corresponding category unit) gives the discriminant function $net_j = \max_k(net_k)$, the Parzen window estimate of the underlying distribution. The $\max_k(net_k)$ operation gives the desired category for the test point.

2 PQ Disturbances Signals Analysis Based on S-Transform

The PQ disturbances signals are complicated, and they are difficult to obtain. The researchers normally use Matlab to simulate the signals and get the test simples [2-8]. 8 types of PQ disturbances include voltage swell, voltage sag, voltage interruption, voltage flicker, harmonic, harmonic with swell, harmonic with sag and voltage transient are simulated by Matlab 7.0 and the simulated simples are analyzed by S-transform aim to extract the feature vector to train and test the PNN classifier. The simulation equations of disturbances are referring to [6, 12] and the sampling rate is 3.2 kHz.

2.1 PQ Disturbances Signals Analysis Based on S-Transform

We show voltage sag and its analysis process in Fig.1 to represent the processes of disturbances analysis.

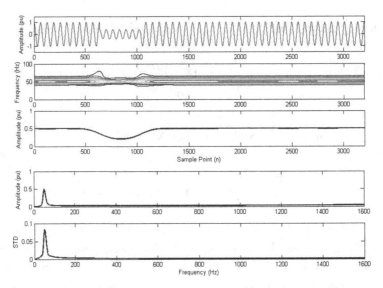

Fig. 1. Voltage sag and its S-transform analysis

Fig.1 is composed by 5 parts. The 1st part shows the plot of the original voltage sag signal. The 2nd part is called time–frequency contour, which is normalized frequency versus time of S-matrix. The 3rd part is called the time–maximum amplitude plot, which is maximum amplitude versus time by searching columns of STA at every frequency. The 4th part is called the frequency–maximum amplitude plot, which is maximum amplitude versus normalized frequency by searching rows of STA at every frequency. The 5th part is called the frequency-standard deviation plot, which shows the standard deviation versus normalized frequency by searching rows of STA at every frequency.

2.2 Disturbances Feature Extraction Based on S-Transform

There are many disturbance features could be used for PQ disturbances recognition, such as energy, standard deviation, autocorrelation, mean, variance, and normalized values. Based on the feature extraction method using S-Transform, 2-dimensional feature set for training and testing is constructed. The 2 features are defined as follows.

Feature 1 (F1): The amplitude factor (A_f), the feature defines as $A_f = \dfrac{A_{\max} + A_{\min} - 1}{2}$, where $A_{\max}$ is the maximum amplitude value of the signal simple and $A_{\min}$ is the minimum amplitude value, $0 < A_f < 1$.

Feature 2 (F2): The standard deviation (σ_{STD}) at the high frequency range (above 100Hz) of FmA-plot.

Feature 3: The minimum value of the maximum amplitude in each cycle of the disturbance simple.

3   Classifier Design and Simulation Test

The classifier is composed by three steps. Firstly, the original signals are analyzed by the S-transform, and distinguished features are extracted from the S-Matrix. Secondly, the selected features are used to compose the input vector of PNN. At last, the disturbances samples are recognized by PNN.

  In the proposed classifier, the neuron number of PNN's input layer is 3, the neuron number of hidden layer is determined by the number of input samples, and the neuron number of output layer is equal to the number of disturbance types.

  50 samples of each type of disturbances with white noise which the SNR is 50dB are simulated to train the PNN. After training the PNN classifier, 100 samples with different SNR 30dB, 40dB and 50dB are simulated to test the trained classifier. Simulation results are shown in Table 1. According to Table 1, the classifier ability is very high even under strong noise environment.

Table 1. Simulation results

Disturbance type	50dB	40dB	30dB
Voltage sag	100%	99%	93%
Voltage swell	100%	100%	97%
Voltage interruption	100%	98%	94%
Voltage flicker	100%	100%	96%
Voltage transient	99%	97%	89%
Harmonic	100%	98%	90%
Harmonic with swell	100%	100%	92%
Harmonic with sag	100%	99%	91%

4   Conclusion

This paper proposed a new approach to recognize power quality disturbances. The original signals are analyzed by S-transform and 3 features are extracted to recognize different types of disturbances. Benefit from the advantages of probabilistic neural network, the new classifier has good classification efficiency and accuracy. Simulation result shows the new approach is satisfied for high noise environment application.

Acknowledgments. This work was supported in part by the National Science Foundation of China (No.50877017 and No.60972065).

References

1. Arrillaga, J., Bollen, M.H.J., Watson, N.R.: Power quality following deregulation. Proc. IEEE 88(2), 246–261 (2000)
2. Herath, H.M.S.C., Gosbell, V.J., Perera, S.: Power Quality (PQ) Survey Reporting: Discrete Disturbance Limits. IEEE Trans. on Power Delivery 20(2), 851–858 (2005)
3. Heydt, G.T., Fjeld, P.S., Liu, C.C., Pierce, D., Tu, L., Hensley, G.: Applications of the Windowed PFT to Electric Power Quality Assessment. IEEE Trans. on Power Delivery 14(4), 1411–1416 (1999)
4. Dwivedi, U.D., Singh, S.N.: Denoising Techniques With Change-Point Approach for Wavelet-Based Power-Quality Monitoring. IEEE Trans. on Power Delivery 24(3), 1719–1727 (2009)
5. Stockwell, R.G., Mansinha, L., Lowe, R.P.: Localization of the complex spectrum: The S-transform. IEEE Trans. Signal Processing 44, 998–1001 (1996)
6. Uyar, M., Yildirim, S., Gencoglu, M.T.: An expert system based on S-transform and neural network for automatic classification of power quality disturbances. Expert Systems with Applications 36, 5962–5975 (2009)
7. Dash, P.K., Samantaray, S.R.: A novel distance protection scheme using time-frequency analysis and pattern recognition approach. Electrical Power and Energy Systems 29, 129–137 (2007)
8. Samantaray, S.R., Dash, P.K., Panda, G.: Fault classification and location using HS-transform and radial basis function neural network. Electric Power Systems Research 76, 897–905 (2006)
9. Specht, D.F.: Probabilistic neural networks. Neural Netw. 3(1), 109–118 (1990)
10. Mishra, S., Bhende, C.N., Panigrahi, B.K.: Detection and Classification of Power Quality Disturbances Using S-Transform and Probabilistic Neural Network. IEEE Trans. on Power Delivery 23(1), 280–287 (2008)
11. Wang, Y., Li, L., Ni, J., Huang, S.H.: Feature Selection Using Tabu Search With Longterm Memories and Probabilistic Neural Networks. Pattern Recognition Letters 30, 661–670 (2009)
12. Gargoom, A.M., Ertugrul, N., Soong, W.L.: Automatic Classification and Characterization of Power Quality Events. IEEE Trans. on Power Delivery 23(4), 2417–2425 (2008)

A Coordinated Heat and Electricity Dispatching Model for Microgrid Operation via PSO

Li Zhong Xu[1], Guang Ya Yang[2], Zhao Xu[3],
Jacob Østergaard[2], Quan Yuan Jiang[1], and Yi Jia Cao[1]

[1] Department of Electrical Engineering, Zhejiang University, Hangzhou, CO 310027 China
[2] Centre for Electric Technology, Department of Electrical Engineering,
Technical University of Denmark, Kgs. Lyngby DK2800, Denmark
[3] Department of Electrical Engineering, Hong Kong Polytechnic University,
Hunghom, Kowloon, Hong Kong
Lizhoxu@gmail.com, gyy@elektro.dtu.dk

Abstract. This paper proposes an optimization model for interconnected Microgrid with hierarchical control. In addition to operation constraints, network loss and physical limits are addressed in this model. As an important component of Microgrid, detailed combined heat and power (CHP) model is provided. The partial load performance of CHP is given by curve fitting method. Meanwhile, electric heater, which supplies heating via electricity, is considered in the model to improve economy of Microgrid operation. The proposed model is formulated into an mixed integer nonlinear optimization problem (MINLP). As an effective tool of nonlinear optimization, particle swarm optimization (PSO) is employed to optimize the operation schedule to minimize the total operational cost of Microgrid considering the jointly optimization of CHP, electric heater and heat storage. Result shows the availability of the proposed algorithm to the model and methodology.

Keywords: Microgrid, Distributed generation, Combined heat and power, Particle swarm optimization, Optimal operation.

Nomenclature

β_{CHP}	Thermal efficiency of CHP	P_{nDis}	Power output from non-dispatchable generators
η_B	Boiler efficiency		
η_{Pump}	Electric heater efficiency	H_B	Heat output of boiler
η_T	Hourly efficiency decay of heat storage	H_T	Heat storage level of heat storage
		H_{TI}	Heat charging rate to heat storage
η_{TI}	Charging efficiency of heat storage	H_{TO}	Heat discharging rate from heat storage
η_{TO}	Discharging efficiency of heat storage	H_D	Heat demand
P_{CHP}	Active power output of CHP	P_{Grid}	Power exchange from the external distribution network
RD_{CHP}	Ramping down limits of CHP	P_{loss}	Power loss of Microgrid

K. Li et al. (Eds.): LSMS / ICSEE 2010, Part I, CCIS 97, pp. 213–219, 2010.

RU_{CHP}	Ramping up limits of CHP	S_{Bij}	Apparent power flow from bus i
P_{Pump}	Power consumption of electric heater		to bus j
U_i	Voltage amplitude of bus i		

1 Introduction

Distributed generation (DG) has been applied rapidly for its potential in significantly reducing emissions and ultimately the system operational cost [1-2]. To utilize the benefits from the emerging benefits of DG, it is necessary to view the DG and the associated loads as a subsystem which is so-called "Microgrid" [3].

The operation of Microgrid offers distinct advantages to customers and utilities, i.e. improved energy efficiency, minimization of overall energy cost, reduced environmental impact, improvement of reliability and resilience, network operational benefits and more cost efficient electricity infrastructure replacement [4].

There are two control strategies that have been proposed to Microgrid operation, decentralized control and hierarchical (centralized) control. In order to achieve the full benefits from the operation of Microgrid, the hierarchical control is determined which contains three control levels.

In [4], potential research issues with Microgrid are proposed and discussed, including generation technologies, design and operation, protection, and economic problem. Research efforts have been carried out on economic problems [5-9]. In [5], optimal Microgrid operation is performed in hourly based manner. Optimal operation approach is used to evaluate system design of Microgrid in [6]. A linear programming based cost minimization model using unit commitment is developed in [7], where system physical constraints and operation limits are addressed and the optimal dispatch of electricity and heat of 24-hour is performed. However, the CHP efficiency is assumed to be constant which may not be practical in real life application. In [8], a novel Energy Management System (EMS) based on Neural Networks (NN) is developed, which can hourly dispatch generators with the goal of minimizing the energy costs. However, network loss and physical limits of Microgrid have not been considered in the study. A 24-hour optimization model of Microgrid is developed in [9], where operational constraints, network loss and physical limits are included, while the electricity and heat are not coordinately dispatched in this model.

In this paper, a Microgrid operation model is developed considering the coordinated dispatch of electricity and heat, network losses, and physical limits. The models of electric heater and storage are introduced into the Microgrid model to verify the flexibility and economy of Microgrid. A day-ahead optimization algorithm is developed for the proposed Microgrid model to minimize its operational cost under a deregulated environment with respect to technical and economical aspects. The operation constraints include DG ramping limits, storage charge/discharge rates, and DER capability limits. Also, the network loss and physical limits are addressed, which may not be neglectable in Microgrid operation [9]. The proposed optimization model is a nonlinear mixed-integer programming optimization problem which normally cannot be solved efficiency by conventional gradient-based methods. In this paper, the advances of recent member in evolutionary algorithm (EA), particle swarm optimization (PSO), is exploited to solve this problem [10].

The paper is organized as follows. In Section II, the schematic figure is provided which illustrates the proposed Microgrid model, followed by the depiction of the components. Also, the modeling of partial load performance of CHP is detailed. In Section III, the day-ahead optimization model for Microgrid is formulated and described. The fundamentals of PSO are described in Section IV. Simulation results are provided and discussed in Section V considering different scenarios. Conclusion is drawn in Section VI.

2 Microgrid

Microgrid may consist of dispatchable and/or non-dispatchable generators, and electrical and/or thermal storages. They are linked by the low voltage network which connects to the external distribution system through Point of Common Connection (PCC) to import or export electricity.

As an important component, CHP generates electricity and provides recoverable heat simultaneously. Since the operation of CHP has contribution on both electricity and heat, its performance has big impact on the Microgrid economical operation. In the previous work, CHP is normally assumed to run at its rated capacity which may not lead to the most economic operation, or its efficiency is assumed to be constant for different load ratio which is not precise enough to reflect its performance. Here partial load model is considered and the performance is modeled by curve fitting method.

In this study, the Capstone200 micro CHP is taken as an example and its performance curve is shown in Fig.2. With polynomial fitting method, the fitting curve is shown below with the original curve.

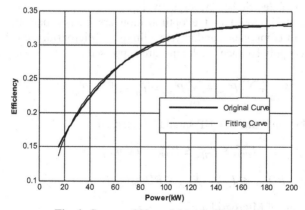

Fig. 1. Capstone200 partial load efficiency

It is assumed that there is more than one unit of C200 installed as one group in the system and the load is shared evenly among all the units. Because of discrete nature of the efficiency curve, integer variables are introduced in the optimization model. In addition to CHP, electric heater is used to convert electricity to heat, which links the electricity and heat in the Microgrid model.

In operation, all controllable components, e.g. CHP, boiler, heat storage and electric heater, are dispatched coordinately to contribute to optimal operation of Microgrid. Different electricity tariffs during the peak and off-peak time are considered in the optimization. Moreover, since the electricity peak demand is in the daytime, while heat peak demand is normally in the night and morning, therefore, CHP may contribute to generate electricity in the daytime to cut the peak, while electric heater can supply heat in the night and morning when electricity tariff is low. Meanwhile, heat storage may aid CHP to store the surplus heat. With this arrangement, the overall operational cost of Microgrid could be optimized.

3   Mathematical Modeling

The objective of optimal operation of Microgrid is to minimize the operational cost with respect to operation and physical constraints. The objective function is presented as [11],

$$
OF = \sum_{t=1}^{24} (c_{Grid}^t P_{Grid}^t + c_{Gas} \sum_{i=1}^{n_{CHP}} f_{CHPi}(P_{CHPi}^t) + c_{Gas} \sum_{i=1}^{n_B} H_{Bi}^t / \eta_{Bi}
$$
$$
+ \sum_{i=1}^{n_{nDis}} c_{nDisi} P_{nDisi}^t + c_T (H_{TI}^t + H_{TO}^t) + c_{Pump} P_{Pump}^t)
$$

(1)

where the first item is electricity exchange cost with external distribution network; the second and third items are fuel consumption costs of CHP and boiler, respectively; and the last three items are the maintenance costs which are proportional to the produced power by renewable resources, charging and discharging of heat storage, and the consumption of electric heater, respectively.

The constraints imposed to the optimization are energy balance, capacity and technical limits of components and physical limits of Microgrid, etc [9]. The electricity demand and supply balance constraints are expressed as

$$
P_{Grid}^t + \sum_{i=1}^{n_{CHP}} P_{CHPi}^t + \sum_{i=1}^{n_{nDis}} P_{nDisi}^t = P_D^t + P_{loss}^t + P_{Pump}^t
$$

(2)

The output and ramp limits of CHP should be satisfied

$$
P_{CHPi}^{\min} \leq P_{CHPi}^t \leq P_{CHPi}^{\max} \quad i \in n_{CHP}
$$

(3)

$$
RD_{CHPi} \leq P_{CHPi}^{t+1} - P_{CHPi}^t \leq RU_{CHPi} \quad i \in n_{CHP}
$$

(4)

The physical limits of Microgrid are

$$
U_i^{\min} \leq U_i^t \leq U_i^{\max} \quad i \in n_{BUS}
$$

(5)

$$S^t_{Bij} < S^{max}_{Bij}, S^t_{Bji} < S^{max}_{Bij} \quad i \in n_{BUS}, j \in n_{BUS}, i \neq j \tag{6}$$

Power exchange with the main grid should be lower than the PCC capacity

$$P^t_{Grid} \leq P^{max}_{Grid} \tag{7}$$

Since the heat generation has to be higher equal than the demand, the constraints are,

$$\sum_{i=1}^{n_{CHP}} H^t_{CHPi} + \sum_{i=1}^{n_B} H^t_{Bi} + \eta_{Pump} P^t_{Pump} + (-H^t_{TI} + \eta_{TO} H^t_{TO}) \geq H^t_D \tag{8}$$

Boiler output constraints are

$$0 \leq H^t_{Bi} \leq H^{max}_{Bi} \quad i \in n_B \tag{9}$$

Electric heater operation constraints

$$0 \leq P^t_{Pump} \leq P^{max}_{Pump} \tag{10}$$

The heat storage operation constraints, charging and discharging limits are expressed

$$H^t_T = (1 - \eta_T) H^{t-1}_T + \eta_{TI} H^t_{TI} - H^t_{TO} \tag{11}$$

$$0 \leq H^t_T \leq H^{max}_T \tag{12}$$

$$0 \leq H^t_{TI} \leq H^{max}_{TI} \tag{13}$$

$$0 \leq H^t_{TO} \leq H^{max}_{TO} \tag{14}$$

4 Particle Swarm Optimization

As mentioned before, the proposed model is a nonlinear mixed-integer programming optimization problem. As a recent member in EA family, particle swarm optimization (PSO) has been proven as an efficient solver in solving complex optimization problems. The algorithm was first proposed in 1995 by James Kennedy and Russell C. Eberhart [10].

PSO is a stochastic, population-based algorithm modeled on swarm intelligence. PSO consists of a population (or swarm) of particles, each of which represents a potential solution. Particles are assigned with random initial positions and fly through problem space with velocities to reach the global optimal solution. In this paper, the advances of PSO are exploited to optimize the operation of the proposed Microgrid model.

5 Case Studies

In this section, a test system is employed to justify the proposed model and method [12]. The model consists of two CHPs with rated power capacity of 800kW and

1000kW respectively, one electric heater, one wind turbine, one boiler, and one heat storage.

In this paper, demand profiles and wind turbine output are assumed to be accurately forecasted beforehand. The electricity and heat demand profiles refer the data from a hotel. Fig. 5 shows demand profiles in January and July.

The operational cost is critically determined by fuel and electricity costs. The electricity tariff is given as time-of-use tariff, which is acquired from a utility. Two-period time-of-use tariff is adopted in this paper, which means that customer is charged at peak rate tariff (0.668RMB/kWh) for any electricity consumed during hour 7-22, and an off-peak tariff (0.288 RMB/kWh) during hour 23-6. In case study, a day-ahead 24 hours optimal scheduling of CHPs, electric heater, boiler and heat storage is implemented.

Simulation results are illustrated in Figure 6, 7. In Fig. 6, the power output of CHPs, power exchange with the external distribution network and the power consumption of

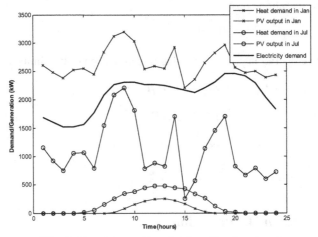

Fig. 2. Electricity and heat load profiles for a hotel

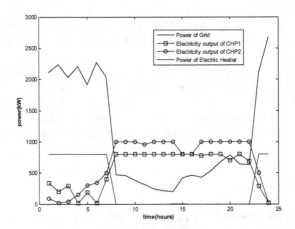

Fig. 3. Electricity power distribution from the main components for winter day

electric heater are shown. It is observed that during hour 1-7 and hour 23-24 when electricity tariff is low, electric heater reaches almost its maximum level, while CHPs are running at low output level. However, during hour 8-22 when electricity tariff is high, outputs of CHPs reaches their upper level, while electric heater remains off. The power exchange curve shows that demand of peak hours is moved to off-peak hours.

6 Conclusion

In this paper, a Microgrid model is proposed and the coordinated heat and electricity optimal operation under day-ahead market environment is achieved via particle swarm optimization. The objective is to minimize the Microgrid operational cost with respect to the physical limits of the grid which is not usually considered in the Microgrid research. Also, the network loss is addressed in the model. The model is applied to a test system for verification of its efficacy.

Future work will focus on considering more constraints, such as unit start-up and shut-down limits, system reserves, reactive power, etc. The consideration on the influence of intermittence of renewable energy on the operation of Microgrid can be another potential for future research.

References

1. Hatziargyriou, N., et al.: Microgrids: an overview of ongoing research, development and demonstration projects. IEEE Power and Energy Magazine 5, 78–94 (2007)
2. Pehnt, M., Cames, M., Fischer, C., Praetorius, B., Schneider, L., Schumacher, K., Voß, J.-P.: Micro Cogeneration: Towards Decentralized Energy Systems. Springer, Berlin (2006)
3. Lasseter, R.H.: MicroGrids. IEEE Power Engineering Society Winter Meeting 1, 305–308 (2002)
4. Lasseter, R.H., Akhil, A., Marnay, C., Stephens, J., Dagle, J., Guttromson, R., Meliopoulous, A., Yinger, R., Eto, J.: The CERTS Microgrid Concept (April 2002)
5. Hernandez-Aramburo, C.A., et al.: Fuel consumption minimization of a microgrid. IEEE Transactions on Industry Applications 41, 673–681 (2005)
6. Zoka, Y., et al.: An economic evaluation for an autonomous independent network of distributed energy resources. Electric Power Systems Research 77, 831–838 (2007)
7. Hawkes, A.D., Leach, M.A.: Modelling high level system design and unit commitment for a microgrid. Applied Energy 86, 1253–1265 (2009)
8. Celli, G., et al.: Optimal participation of a microgrid to the energy market with an intelligent EMS. In: The 7th International Power Engineering Conference IPEC, vol. 2, pp. 663–668 (2005)
9. Mashhour, E., Moghaddas-Tafreshi, S.M.: Integration of distributed energy resources into low voltage grid: A market-based multiperiod optimization model. In: Electric Power Systems Research, vol. 80, pp. 473–480 (2010)
10. Kennedy, J., Eberhart, R.: Particle swarm optimization. In: Proceedings of IEEE International Conference on Neural Networks, vol. 4, pp. 1942–1948 (2005)
11. Siddiqui, F.R., Ghosh, A.S., Stadler, S., Marnay, M., Edwards, C., Edwards, J.L., Marnay, C.: Distributed energy resources with combined heat and power applications. Ernest Orlando Lawrence Berkeley National Laboratory (June 2003)
12. Kersting, W.H.: Radial distribution test feeders. IEEE Transactions on Power Systems 6, 975–985 (1991)

MPPT Strategy of PV System Based on Adaptive Fuzzy PID Algorithm

Jing Hui and Xiaoling Sun

School of IOT engineering, Jiangnan University,
214122 Wuxi, JiangSu, China
xiaoling_sun1985@126.com, jingh@126.com

Abstract. To further improve the control quality of photovoltaic generation MPPT system, dual-mode adaptive fuzzy PID control strategy was proposed in this paper. On the basis of conventional fuzzy tracking algorithm, the principle of control algorithm was analyzed and the control system was designed. The results show that dual-mode control algorithms can quickly sense the changes in the external environment, and track the maximum power point rapidly. At the same time, oscillation phenomenon near the MPP is eliminated effectively. The proposed MPPT system represents good stability, accuracy and rapidity.

Keywords: photovoltaic generation, MPPT, adaptive fuzzy control, PID, dual-mode control.

1 Introduction

As the clean renewable energy, the solar power has got sustained development and been widely used. There is a growing concern about photovoltaic power generation which is one of the main ways to use solar. The output properties of photovoltaic cells are greatly affected by the external environment both the battery surface temperature and sunlight intensity change can lead to great changes of the output characteristics, and then reduce the conversion efficiency of photovoltaic cells. At present, the conversion efficiency of monocrystalline silicon cells is generally 12%-18%, and that of polycrystalline silicon is only 12%-17%. Therefore, an efficient Maximum Power Point Tracking control algorithm has an important significance in improving the efficiency of photovoltaic system, and in photovoltaic generation industry as a whole.

Maximum power point tracking is based on load line adjustment under varying atmospheric and load conditions by searching for an optimal equivalent output resistance of the PV module, and many tracking control strategies have been proposed. Conventional MPPT control algorithms such as the disturbance observation method (commonly known as hill-climbing) and the incremental conductance method are simple and easy to implement. However, it still has many disadvantages such as high cost, complexity, instability, the oscillations near maximum power point causing power loss, poor adaptability to the external environment, low robustness of the system and so on[1][2]. In view of the problems in conventional MPPT control algorithms, drawing on the advantages and disadvantages of each control algorithm in references [2], [3] and [4], a dual-mode

K. Li et al. (Eds.): LSMS / ICSEE 2010, Part I, CCIS 97, pp. 220–228, 2010.

adaptive fuzzy PID control strategy was proposed, the algorithm theory was introduced in detail and the system model was established. Experimental results show that the dual-mode control algorithm can significantly reduce oscillations near the maximum power point of the conventional fuzzy control. It improves the system's stability, at the same time; the introduction of adaptive control enhances environmental adaptability of fuzzy PID control algorithm. The algorithm has good robustness and control precision, thus achieves the unity of rapidity and accuracy.

2 Theory of MPPT of Photovoltaic Generation System

Solar photovoltaic generation system is a power generation system which takes use of photovoltaic effect of semiconductor materials, and directly converts solar radiation to electricity. As can be seen from Figure 1, there is a maximum power point of the photovoltaic cells' output power. The formula (1) shows the relationship.

$$p_m = I_m * V_m \tag{1}$$

At the same time, the maximum output power of photovoltaic cells changes with the changes of the external environment. To improve the conversion efficiency of photovoltaic cells, a maximum power point tracking method is required.

The process of maximum power point tracking is essentially a self-optimizing process. Presently, the hill-climbing and incremental conductance methods are commonly used in MPPT control. The former has a simple structure, little perturbation parameters, and relatively good tracking efficiency; however, the tracking fluctuation results in power loss. The latter can quickly track the changes of the maximum power point caused by changes of light intensity and has a better tracking effect, but the hardware implementation is very difficult. Fuzzy logic control does not require modulating output voltage so as to avoid some of the power loss, but its control rules can not be amended according to changes in the external environment, and self-tune parameters to make the system stable at the maximum power point. Compared with advantages and disadvantages of these control methods, integrating fuzzy control and classical PID control, an adaptive fuzzy PID control algorithm with both accuracy and rapidity is raised in this paper.

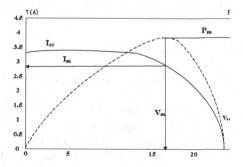

Fig. 1. I-V characteristics of photovoltaic array

3 Principle of Adaptive Fuzzy PID Control

In order to overcome the shortcomings of conventional fuzzy control rules and give full play to the performance of PV cells, the adaptive fuzzy control algorithm with online system parameters self-tuning was proposed. Depending on real-time data to achieve the automatic adjustment and improvement of fuzzy control rules in the control process, it can achieve a good precision, at the same time, in order to reduce the oscillation near the maximum power point, and reduce the system's fluctuations and energy loss; the traditional PID control method was introduced. Adaptive fuzzy PID controller was constructed after taking advantage of adaptive fuzzy control with the features self-tuning, and integrating the stability and rapidity of traditional PID control [3]. The controller can achieve parameters auto-tuning, enhance the stability of the system, reduce energy loss, and then improve energy conversion efficiency.

Principle of the dual-mode adaptive fuzzy PID control is shown in Figure 2. First, voltage and power values were collected. Power values determined its work area, and then adopted different work orders to realize tracking control according to different workspaces. In the context of large deviations, fuzzy adaptive control was used for rapid response, and the accuracy adjustment within the scope of a small deviation used conventional PID control. The switching function $k_{(ep)}$ determined the switch between the two kinds of control mode. K1 and K2 was the switch sets of adaptive fuzzy controller, and the value of the selection was decided by different environmental conditions and field experience. Conventional PID control was taken when meeting the condition '$k_1 < \Delta p / \Delta D < k_2$', in addition, adaptive fuzzy control was used [4].

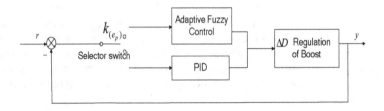

Fig. 2. Dual-mode adaptive fuzzy PID control principle diagram

4 Algorithm and Controller Design

In the photovoltaic generation system, the maximum power point tracking speed and tracking accuracy is the key factors of control system, and these factors directly relate to the regulation step of the system. When the system operating point is away from the maximum power point, the tracking speed must be accelerated, that is, to increase regulation step; when the operating point is near the maximum power point, in order to improve system's tracking accuracy and stability, step length must be properly reduced to avoid system oscillation [5]. Therefore, the maximum power point on both sides needs to take different duty cycle to achieve maximum power tracking of photo-voltaic cells.

4.1 Adaptive Fuzzy Controller Design

According to the characteristics of PV systems, the initial fuzzy controller chose the triangular shape as membership function shape. When the curve closer to the origin (smaller errors), the curve became steeper (higher resolution); when it farther away from the origin, the curve became more slowly. According to the characteristics of photovoltaic cells, it could be drawn that: when the point was more distant from the maximum power point, a larger step size should be taken to speed up the tracking speed. In contrast, a smaller step size should be used to reduce the search loss. If the changes of temperature, sunlight intensity and other factors leaded to significant changes of photovoltaic power system, the system must respond quickly.

Two inputs of adaptive fuzzy controller is error e and error change Δe, as the formula (1) shows. Output ΔD is the change amount of duty cycle of the MPPT circuit switching device.

$$e(k) = \frac{p(k) - p(k-1)}{v(k) - v(k-1)},$$

$$\Delta e(k) = e(k) - e(k-1) \tag{2}$$

As is shown in fig 3, adaptive fuzzy controller is based on the fuzzy control by 3 additional functional blocks, respectively performance calculation (identification devices), decision-making body, and the control rules to amend institutions.

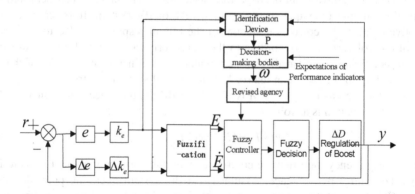

Fig. 3. Block diagram of adaptive fuzzy control system

The actual response for each sample of identification device can be obtained by monitoring $e(kT)$ and $\Delta e(kT)$. After comparing the actual response with the hope response, the output needing corrected was indicated, that is, to determine the correction amount of the output response so as to provide the information for amendments of control rules. The data in Table 1 gave hope response. Zero elements indicated that the state didn't require correction. The value of non-zero elements not only took into account the distance deviated from the set value, but also the speeds that tended to set value and left away from the set value.

Table 1. Decision table of performance measurement

e \ Δe		Close to the maximum power point							Away from the maximum power point						
		-6	-5	-4	-3	-2	-1	-0	+0	+1	+2	+3	+4	+5	+6
The left of maximum power point	-6	0	1	2	2	2	2	6	6	6	6	6	6	6	6
	-5	0	0	1	3	3	3	5	5	5	5	5	5	6	6
	-4	0	0	0	3	3	3	5	4	4	5	5	5	5	6
	-3	0	0	0	2	2	1	4	3	4	4	5	5	5	6
	-2	0	0	0	0	1	1	3	2	3	3	4	4	5	5
	-1	0	0	0	0	0	1	2	1	2	2	3	3	4	5
	-0	0	0	0	0	0	0	0	0	0	0	1	2	3	4
The right of maximum power point	+0	0	0	0	0	0	0	0	0	0	0	-1	-2	-3	-4
	+1	0	0	0	0	-1	-1	-2	-1	-2	-2	-2	-3	-3	-4
	+2	0	0	0	0	-2	-3	-4	-4	-4	-4	-5	-5	-5	-6
	+3	0	0	0	-1	-2	-3	-4	-4	-5	-5	-5	-5	-5	-6
	+4	0	0	0	-3	-3	-3	-4	-4	-4	-5	-5	-5	-5	-6
	+5	0	0	0	-3	-3	-3	-5	-5	-5	-5	-5	-5	-6	-6
	+6	0	0	-1	-2	-2	-3	-5	-6	-6	-6	-6	-6	-6	-6

Through the above-mentioned performance measurement has been achieved the correction amount of output response for MPPT of PV system. In order to achieve adaptive control, the correction amount of the output response needed to convert to that of control variables. According to the characteristics of control object, an incremental model was established. That is, to calculate the incremental model of the object according to the Jacobean matrix J of the control system's output to the input. In that way, the incremental model M is $M = TJ$, and M is the object state function. The amount of correction is as follows.

$$\Delta u(kT) = M^{-1}\Delta y(kT) \tag{3}$$

Amendment agency modified the control rules by the correction amount so as to improve the control performance. Assumed that the system operating point deviation from MPP due to the external environment change in the cap d sub-sampling when the point of error, the error rate of change, and the control input was respectively $e(kT - dT)$, $\Delta e(kT - dT)$, $u(kT - dT)$. Based on the control correction amount of calculation, control input should be taken as

$$v(kT - dT) = u(kT - dT) + \Delta u(kT) \tag{4}$$

In order to obtain revised strategy, the corresponding fuzzy set was constructed according to the amount of corresponding domain [6], and replaced the old implication (5)with the new implication(6).

$$E(kT-dT) \rightarrow \dot{E}(kT-dT) \rightarrow V(kT-dT) \tag{5}$$

$$E(kT-dT) \rightarrow \dot{E}(kT-dT) \rightarrow U(kT-dT) \tag{6}$$

And then the corrected fuzzy rules was If $e(kT-dT)$ is $E(kT-dT)$ and $\Delta e(kT-dT)$ is $\dot{E}(kT-dT)$ then $u(kT-dT)$ is $v(kT-dT)$.

4.2 Dual-Mode Controller Design

Adaptive fuzzy control algorithm is better than conventional fuzzy one with better precision and self-correcting capability, however the oscillation near the maximum power point leading to the larger energy loss has not got resolved satisfactorily, as a result affects the stability and conversion efficiency of overall system. To solve this problem, PID control with stability and rapidity was introduced on the basis of fuzzy algorithm. The switch between the two kinds of control mode was decided by the switch function $k_{(ep)}$ [7]. Integrate the advantages of intelligent algorithm and classical algorithms to achieve good control performance.

5   Experimental Set Up and Results

Experimental device mainly consisted of photovoltaic modules, Boost circuit and a self-adaptive fuzzy PID controller as fig 4 showed. PV module parameters were as follows: peak power p_{mp} was 9W, open-circuit voltage V_{OC} was 21V, short-circuit current I_{SC} was 0.6A, maximum power voltage V_{mp} was 16.8V, maximum power current I_{mp} was 0.54A, and NOCT (normal cell operating temperature) T_{noct} was $50°C$.

The core of control system was adaptive fuzzy PID controller, and it was achieved by TMS320LF2812DSP made in TI. Output voltage and output current signals detected by PV module were sent to the controller. After processing the voltage and current signals, the regulation amount of the main switch duty cycle of Boost circuit was obtained eventually so as to control switch changes of Boost circuit, the process didn't stop until the system worked in the MPP.

Comparing with the experimental results of fig 5(a) and 5(b), it can be drawn that the improved dual-mode control algorithm can effectively reduce the oscillation near the maximum power point compared to the only adaptive fuzzy algorithm, and output holds stable. Thus the new system increases the output power of photovoltaic generation system.

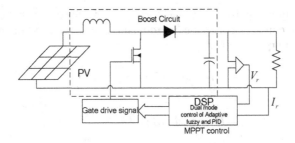

Fig. 4. Schematic of PV MPPT system

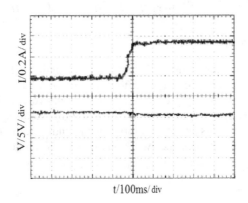

Fig. 5(a). The experimental waveforms of dual-mode adaptive fuzzy PID MPPT algorithm

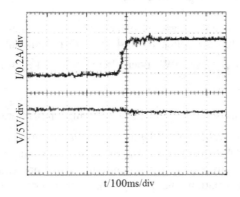

Fig. 5(b). The experimental waveforms of adaptive fuzzy MPPT algorithm

Output power under the two algorithms was drawn by referring references [8]. The comparison result shows that approximately 5% more output power has been extracted from the PV arrays based on the control algorithm proposed in the text and thus higher conversion efficiency was achieved.

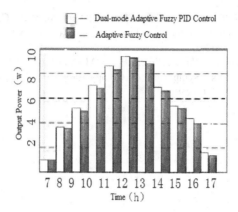

Fig. 6. Output power generated by the systems of adaptive fuzzy and dual-mode adaptive fuzzy PID

6 Conclusions

Dual-mode adaptive fuzzy PID control algorithm was proposed in this paper after integrating the advantages and disadvantages of adaptive fuzzy control and conventional PID control. As can be seen from the experimental results, the new control algorithm can effectively reduce the oscillation near the maximum power point, decrease power loss and improve conversion efficiency of photovoltaic cells compared with the conventional fuzzy control. In the case of mutations of light intensity (to cover photovoltaic panels in the experiment), the system is able to find a new maximum power point quickly to maintain system stability, and improve the robustness and accuracy of MPPT control system, at the same time enhance the stability of the tracking system.

References

1. Otieno, C.A., Nyakoe, G.N., Wekesa, C.W.: A Neural Fuzzy Based Maximum Power Point Tracker for a Photovoltaic System. In: IEEE Africon, Nairobi, 1–6 (2009)
2. Zhengqi, Y.: Study of MPPT Technique Based on Fuzzy Logic Control in PV System. D. Huazhong University of Science and Technology, Wuhan (2007)
3. Xinghong, Q., Bijun, W.: Fuzzy / PID dual-mode MPPT control in the application of PV power generation. J. Electric Power Automation Equipment, 93–94 (2008)
4. Cao, z., Xiangning, H.: Maximum Power Point Tracking by Using Asymmetric Fuzzy Control Combined With PID for Photovoltaic Energy Generation System. J. Transactions of China Electro Technical Society, 74–75 (2005)
5. Yongjun, L., Pin, W.: Adaptive fuzzy algorithm in the application of PV MPPT systems. J. Acta Energiae Solaris Sinica, 658–660 (2008)
6. Yanjun, L., Ke, Z.: Adaptive Control Theory and Applications. Northwestern Polytechnical University Press, Xian (2005)

7. Yan, X., Sun, J.: Self-adaptive tuning of fuzzy PID control of PV Grid-connected inverter. In: Sixth International Conference on Fuzzy Systems and Knowledge Discovery, Tianjin, vol. 4, pp. 160–162 (2009)
8. Guohui, Z., Qizhong, L.: An Intelligent Fuzzy Method for MPPT of Photovoltaic arrays. In: Second International Symposium on Computational Intelligence and Design, Changsha, vol. 2, pp. 356–359 (2009)

Optimized Approach to Architecture Thermal Comfort in Hot Summer and Warm Winter Zone

Xianfeng Huang[1,*] and Yimin Lu[2]

[1] Education Ministry Key Laboratory of Disaster Prevention and Structural Safety &
College of Civil Engineering and Architecture, Guangxi University,
100 Daxue Rd.,530004 Nanning, China
seulee@163.com
[2] College of Electrical Engineering, Guangxi University,
100 Daxue Rd., 530004 Nanning, China
aulym@163.com

Abstract. According to the Fanger thermal comfort equations and the advantage of artificial immune algorithm in solving combinatorial optimization to engineering problems, the artificial immune algorithm has been applied to the thermal design and optimization of parameters for the HVAC (Heating, Ventilation, and Air Conditioner) within a building. Then, aiming at climate characteristics of high temperature and humidity in hot summer and warm winter zone, the interior thermal comfort objective function is deduced. Then, the preferable range of indoor air temperatures and air velocities which meet the thermal comfort requirements under different activities and humidity is obtained by simulation. Therefore, the relationship between indoor thermal comfort and energy saving to a building is also managed to reveal. Furthermore, it is evidence that the proposed method in this paper should be employed in the criteria for thermal design to a building and control of an air conditioning system.

Keywords: Artificial Immune Algorithm; Thermal Comfort; Optimization; Hot Summer and Warm Winter Zone.

1 Introduction

Generally, in hot summer and warm winter zone, the indoor air temperature exceeds 28℃ which is maximum of thermal comfort temperature in a chamber with no air condition in summer. For the sake of satisfaction with the thermal environment, the requirements of air condition for chamber result in the increasing of energy consumption rapidly. Consequently, summer is a major season for thermal comfort and energy saving design to a building in this zone.

The architectural thermal comfort is influenced by many parameters, which mainly include factors such as air temperature, humidity, air velocity, radian temperature of

---

* Project Supported by Foundation of Education Ministry Key Laboratory of Disaster Prevention and Structural Safety (2009TMZR003), National Natural Science Foundation of China (NSFC) (60774024).

K. Li et al. (Eds.): LSMS / ICSEE 2010, Part I, CCIS 97, pp. 229–237, 2010.
© Springer-Verlag Berlin Heidelberg 2010

each interior surface, thermal resistant of clothes and metabolism etc. According to the thermal equilibrium equation in a steady thermal state, Professor P.O. Fanger proposed the Predicted Mean Vote (PMV) and Predicted Percentage Dissatisfied (PPD) to indicate people's thermal sensation and dissatisfying to a certain thermal environment. The PMV and PPD are expressed by complicated, coupled and nonlinear equations which solutions are hardly determined by regular ways, and the computation is a considerable tough job when it involves a large number of design variables.

2 Indoor Air Temperature, Humidity and Air Flow in Summer

In summer, the indoor air temperature, humidity and air movement are the significant parameters which influence indoor thermal comfort and energy saving within a whole building directly. Indoor air temperature affects human thermal comfort sensitively; humidity also has obvious effect on the heat loss on body skin, skin witteness, and evaporating from the ambient environment etc. If indoor humidity keeps within a domain of 40%-70%, it will not hinder the sweat from evaporating, and influence heat dissipation rate and the skin surface temperature evidently; and, air movement can accelerate heat dissipation from human body. Out of climate characters in hot summer and warm winter zone, favorable ventilation, i.e. evenness of airflow, effective drafts feelings, posses an indubitable importance for keeping interior thermal comfort [1-3]. Therefore, the optimal design of the thermal comfort must satisfy multiple constraints.

3 PMV Thermal Comfort Model

The PMV thermal comfort model is based on extensive experiments involving over many subjects exposed to well-controlled indoor environment, and can predict the thermal sensation as a function of indoor air temperature, mean radiant temperature, air velocity, humidity, human activity and clothing[2]. The International Standards Organization (ISO 7730) adopts what Prof. Fanger developed PMV-PPD significant index system as criterion for evaluating and measuring the indoor thermal environment. PMV is defined as

$$\text{PMV} = (0.303e^{-0.036M} + 0.0275)\{M - W - 3.05[5.733 - 0.007(M - W) - p_a] - 0.42(M - W - 58.15) - 0.0173M(5.867 - p_a) - 0.0014M(34 - t_a) - Q\} , \tag{1}$$

where M is the heat generated by metabolism, depending on the activity, W/m^2; W is the work done by human activity, W; p_a is ambient vapor partial pressure, represents relativity humanity, kPa; t_a: ambient air temperature, ℃; Q is the apparent heat load acting on the human body, W:

$$Q = 3.96 \times 10^{-8} f_{cl}[(t_{cl} + 273)^4 - (t_r + 273)^4] + f_{cl}h_c(t_{cl} - t_a) , \tag{2}$$

$$t_{cl} = 35.7 - 0.028(M - W) - 0.155 I_{cl}Q , \tag{3}$$

$$h_c = \begin{cases} 2.38(t_{cl} - t_a)^{0.25} & 2.38(t_{cl} - t_a)^{0.25} > 12.1\sqrt{v} \\ 12.1\sqrt{v} & 2.38(t_{cl} - t_a)^{0.25} < 12.1\sqrt{v} \end{cases} , \tag{4}$$

where t_{cl} is the temperature of clothing surface, °C; h_c is the convective heat transfer coefficient, W/(m²·K); v is air velocity, m/s; I_{cl} is clothing insulation, clo; f_{cl} is clothing factor accounting for the relative increase in the clothed body surface over the unclothed. The relationship between I_{cl} and f_{cl} is $f_{cl} = 1 + 0.25I_{cl}$, as to the garments during summer $I_{cl} = 0.5$clo. t_r is the average radiance temperature within a whole room, °C

$$t_r = \frac{\sum_{j=1}^{n} S_j \theta_{i,j}}{\sum_{j=1}^{n} S_j},$$ (5)

where S_j are surface area of each wall, ceiling and floor within a room, m². After determination of PMV index, PPD value can be obtained by the equation below

$$PPD = 100 - 95e^{-(0.03353PMV^4 + 0.2179PMV^2)}.$$ (6)

PMV-index predicts the subjective ratings of the environment in a group of people, and is divided into 7 evaluated thermal sensation scales. Thermal neutral (comfort) is presented as PMV = 0; warm (or hot) feeling is correspond to the inequality PMV > 0; whereas, cool (or cold) feel refers to the inequality PMV < 0. The commendatory values of thermal comfort by ISO7730 are -0.5 ≤ PMV ≤ 0.5 and PPD ≤ 10%. The solutions of PMV function agrees well with thermal sensation in building with HAVC in summer. So, satisfying the inequality 0≤ PMV≤ 0.5 will maintain indoor thermal comfort and reduce HVAC energy consumption theoretically, while refrigeration in progressing.

Owing to iteration of the thermal comfort equations, it is impracticable to achieve the solutions of PMV and PPD at first hand. Furthermore, under satisfying thermal comfort and energy saving, it exists also difficulty to figure out each preferable variation range of parameters for combinatorial optimization. It should be pointed out that different combination of parameters incurs dissimilar people's thermal sensation, so the thermal parameters should be comprehensively analyzed. Many optimization algorithms have been proposed for the nonlinear programming solution. While each technique has its own advantages and disadvantages, including various degrees of efficiency, none of them are suitable for all problems. Therefore, the option of an optimization algorithm for thermal comfort depends on specific engineering problem, the search method should have high directional flexibility, and convergence to a global optimum should be guaranteed. Due to the memory cells that guarantee the fast convergence toward the global optimum and affinity calculation routine to embody the diversity of the real immune system, these differences between artificial immune algorithm and the other probabilistic optimization algorithms assure artificial immune algorithm is employed to solve a variety kind of optimization practical engineering problems and can be developed to the inverse prediction of thermal comfort[4].

4 Optimization Method

The immune system is a high mammal defense system against foreign invader. The immune system has mechanisms that recombine the gene to copy with the invading antigens, produced the antibodies and exclude the antigens, and has fundamental ability to produce new types of antibody or to find the best fitted antibody which is able to attack the antigen invading into the body. Artificial immune algorithms (AIA) are biologically inspired optimization techniques for parallel globe search, which imitate immune systems of an organism to solve the combinatorial optimization problems[5-8]. By the description of AIA, the objectives and constraints of an optimization problem are thought as antigens, while on the other hand, the feasible solutions of the problem are regarded as antibodies. By some efficient operations, such as recognizing, selection, mutation and crossover operations in antibodies at each iterative, the optimized solution to the problem is obtained. The flow chart of immune algorithm is shown as Fig. 1.

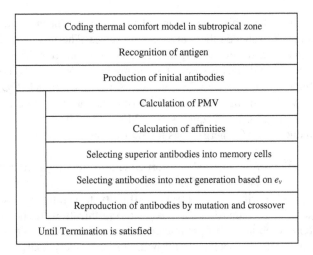

Fig. 1. Flow chart of immune algorithm

The procedure of antigens acceptance is defined as solving for the objective function to a problem. The value of PMV is determined by six factors, of them, both of the indoor air temperature and the air velocity are the significant parameters for a building thermal design and HVAC control. It is shown that there is a wee effect on human thermal sensation within a certain indoor environment, with the assumption that the air ambient temperature equals to the average radiant temperature $t = t_a = t_r$. Then the objective function is defined as

$$\text{opt} = |\text{PMV}(t, v)| . \tag{7}$$

While the value of the objective function arrives at minimum, the amount of PMV closes to zero which means attaining thermal comfort. Two variables of the objective function are indoor air temperature t and air velocity v which are constitute of antibodies' coding string. Each antibody corresponds to a set of the indoor environment parameters (t, v) that are candidates for the solutions of (7). Additionally, the solutions to the thermal comfort inverse prediction are concerned with the calculation of successive functions other than with the abstract symbols. The real number coding, thus, may close to the reality of problems. Corresponding to some limitation and specification of thermal environment in summer, supposed that t and v are confined within $[10, 30]°C$ and $[0, 0.25]$ m/s respectively. Each antibody, therefore, can be expressed as real strings that involve the two variables. The initial antibody population is generated within definition domain of the independent variables randomly.

In order to measure similarity between antibodies or antigens, two forms of affinity are calculated. One is the affinity between two antibodies, which means the diversity of the immune system. Let α and β be two antibodies in the immune system. Then the affinity between α and β can be defined as

$$ay_{\alpha,\beta} = 1/(1+ dist_{\alpha,\beta}) , \tag{8}$$

where $dist_{\alpha,\beta}$ is the weighted distance between α and β: $dist_{\alpha,\beta}=|t_{\alpha}-t_{\beta}|+k|v_{\alpha}-v_{\beta}|$, k is a weighted value. The maximum value of affinity is 1. While another affinity between antibody α and antigen is expressed as

$$ax_{\alpha} = 1/(1+ opt_{\alpha}) , \tag{9}$$

where opt_{α} is the objective function of antibody α. The value of affinity will equal to 1, while an antibody matches an antigen completely, i.e. the larger affinity value indicates stronger affinity. The probability of an antibody α will be selected to the next generation is determined by its expectation rate e_{α}

$$e_{\alpha} = ax_{\alpha}/ c_{\alpha}, \tag{10}$$

where c_{α} is the density of antibody α. Suppose an immune system is composed of a certain number of antibodies n, c_{α} is calculated by

$$c_{\alpha} = \frac{1}{n}\sum_{\beta=1}^{n} ac_{\alpha,\beta} , \tag{11}$$

in which

$$ac_{\alpha,\beta} = \begin{cases} 1 & ay_{\alpha,\beta} \geq Tac_1 \\ 0 & otherwise \end{cases} . \tag{12}$$

Tac_1 is a defined threshold beforehand. From (10) it can be seen that the antibodies with high affinity to the antigen are promoted, while the antibodies with high density are suppressed. Thus the diversity of the system which adapt to different antigen can be maintained through this adjusting mechanism.

To accelerate convergence accessible toward the optimal solution, some antibodies with high affinity to the antigen are opted into the memory cells and may be used to produce some of the initial antibodies. In the inverse prediction of thermal comfort, the solutions that content inequality $0 \leq PMV \leq 0.5$ and have an acceptable deviation are required. So the antibodies that satisfied with the requirement at each process will be stored into the memory cells. Furthermore, a suppressor works to eliminate the surplus of solution candidates when the concentration c_{α} of antibody exceeds over a threshold level *Tc*. This process can be regarded as keeping appropriate solution(s) as memory for the next search step.

The reproduction of the next generation is performed within the survival antibodies by using mutation and crossover operators.

5 Simulation

Three design variables were selected for the optimization study, including indoor air temperature, indoor humilities, and indoor air flow. Due to the effect of humidity on the thermal comfort in this zone, the domain of indoor air temperature and air flow parameters can be determined by setting different humidity. According to ISO7730 and higher humidity (70% more) in hot summer and warm winter zone in summer, with diverse indoor humidity (Partial vapor pressure: 2.5kPa, 2.8kPa and 3.0kPa respectively) and under different activities (Metabolic Rate: seated relaxed $M=58.2W/m^2$, standing relaxed $M=70W/m^2$, walking or light activity $M=93W/m^2$), the thermal comfort optimization are progressed in this paper.

The AIA was set with a population size of $n=100$, crossover rate $p_c=0.9$, mutation rate $p_m=0.1$, calculation time 50. The best affinity between antibody and antigen in each generation is as shown in Fig. 2. When affinity between antibody and antigen is approached to 1, the deviation between expected and predicted of sound insulation is close to 0.

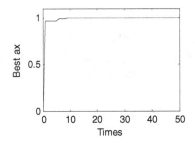

Fig. 2. The best affinity between antibody and antigen at each generation

Owing to the arbitrary initialization, the initial generated antibodies distribute uniformly in the whole parameter space, and the simulation data exhibits a strip, which corresponds to the comfort zone, while the last generation antibodies with diversity distribute around comfort area. The antibodies in memory cell that contents the thermal comfort condition $0 \leq PMV \leq 0.5$ is close to the optimized antibodies, i.e. the optimized solution to indoor air temperature and air velocity, are shown in Fig.3, 4, 5. Under the condition of thermal comfort and minimum energy consumption, these Figures reveal that optimized values indoor air temperature and air flow.

The thermal comfort simulations were feasible conducted for paralleled search a comfort extent. While partial vapor pressure is 2.5kPa corresponds to a relative low indoor humidity which benefits to comfort sensation, the distribution of antibodies (with the symbol "x") with different activities is shown in Fig. 3. While partial vapor

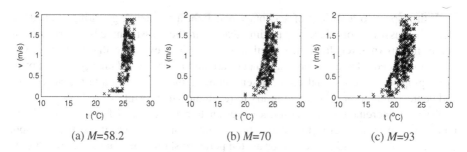

Fig. 3. While p_a=2.5kPa and thermal comfort, the optimized combination of interior temperature and air velocity (Antibodies in memory cell)

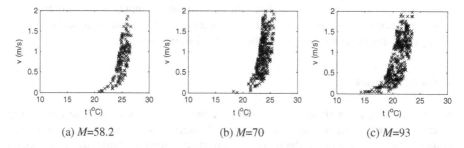

Fig. 4. While p_a=2.8kPa and thermal comfort, the optimized combination of interior temperature and air velocity (Antibodies in memory cell)

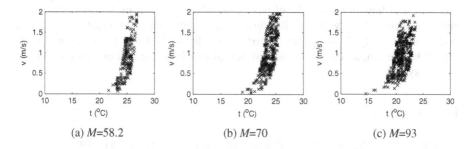

Fig. 5. While p_a=3.0kPa and thermal comfort, the optimized combination of interior temperature and air velocity (Antibodies in memory cell)

pressure is 2.8kPa corresponds to a moderate indoor humidity, the distribution of antibodies with different activities is shown in Fig. 4. While partial vapor pressure is 3.0kPa corresponds to a relative high indoor humidity which baffles comfort feeling, the distribution of antibodies with different activities is shown in Fig. 5.

In Fig. 3, 4 and 5, under the condition of the thermal comfort, it hints that the indoor temperature should reduce to meet the increasing metabolic rate. With raise of the indoor air temperature, the indoor air velocity should go up to attain the thermal comfort feeling. On the other hand, Even at a lower relative humidity, thermal comfort sensation may be obtain at a higher air temperature.

With respect to a variety of sedentary and light activities, e.g. sitting students or clerks in classroom or office, or standing teacher in the classroom, customers walking gently in supermarket, from the simulation result, it is evidence that higher activity levels may need a lower indoor air temperature and more HVAC energy consumption to keep up with the thermal comfort sensation; or may also promote the ventilation and depress indoor air relative humidity to reach the same result.

With the increment of humidity, a lower air temperature will make a HVAC system consume more energy to balance the thermal sensation, e.g. if the partial vapor pressure is above 2.5kPa, the indoor air temperature should be limited below 25℃ to reach the comfort extent. In the place of the light activity, e.g. classroom or office the moderate indoor air flow will keep thermal comfort at higher air temperature; moreover, in the place with slightly higher activity levels, such as supermarket or railway station, the thermal comfort sensation will be achieve by suppressing indoor air humidity and accelerating air flow. Thus the combinatorial optimization between thermal comfort and energy saving might be achieved simultaneously.

6 Conclusions

In the light of Fanger thermal equations and the advantages of AIA in optimizing engineering problems, the preferable range of interior air temperatures and air velocities which accommodate the thermal comfort requirements under different activities is obtained by simulation. Consequently, within the comfortable extent of thermal parameters, a key of thermal design to a building is reasonable combination of indoor air flow, temperature and humidity. It is further, proved that the higher air humidity and human activity level have a significant impact on thermal uncomfortable sensation, which will be felt intensively, as the gentle increment of the air humidity and activity. It is evidence, hereby, that the method that proposed in this paper should be employed in the thermal design to a building and energy control of a HVAC system.

References

1. Fanger, P.O.: Thermal Comfort. Danish Technical Press, Copenhagen (1970)
2. Fanger, P.O., Toftum, J.: Extension of the PMV Model to Non-air-conditioned Buildings in Warm Climates. Energy and Building 34, 534–536 (2002)
3. Wang, H.Q., Huang, C.H., Liu, Z.Q., et al.: Dynamic Evaluation of Thermal Comfort Environment of Air-conditioned Buildings. Building and Environment 41(11), 1522–1529 (2006)
4. Wang, T., Li, S., Nutt, S.R.: Optimal Design of Acoustical Sandwich Panels with a Genetic Algorithm. Applied Acoustics 70(3), 416–425 (2009)
5. Lu, Y.M., Huang, X.F., Mao, Z.Y., et al.: Optimization Thermal Comfort Model Using Artificial Immune Algorithm. Journal of Harbin Institute of Technology 37(3), 355–358 (2005)

6. Hugues, B.: The Endogenous Double Plasticity of the Immune Network and the Inspiration to Be Drawn for Engineering Artifacts. In: Dasgupta, D. (ed.) Artificial Immune Systems and Their Applications, pp. 22–44. Springer, Berlin (1999)
7. Toyoo, F., Kazuyuki, K., Makoto, T.: Parallel Search for Multi-modal Function Optimization with Diversity and Learning of Immune Algorithm. In: Dasgupta, D. (ed.) Artificial Immune Systems and Their Applications, pp. 210–220. Springer, Berlin (1999)
8. Chun, J.S., Kim, M.K., Jung, H.K.: Shape Optimization of Electromagnetic Devices Using Immune Algorithm. IEEE Transactions on Magnetic 33(2), 1876–1879 (1997)

The Application of Computational Fluid Dynamics (CFD) in HVAC Education

Jiafang Song[*] and Xinyu Li

Department of Building Environment and Equipment Engineering,
Tianjin Polytechnic University, Tianjin, China, 300160
songjiafang@tjpu.edu.cn

Abstract. In this paper, we show the application of CFD in HVAC education. In the course, it was conducted using Fluent CFD software to improve the ventilation performance in one classroom. The CFD approach provided simulation results with different numbers of fans installed in the classroom. In the CFD simulations, the models with fans or without fans were created. By comparing the different cases simulated by Fluent software, it was found that installing fans could improve the ventilation performance in the classroom effectively. It also can be seen from the CFD simulation results that the numbers of the fans and installing positions of the fans can affect the ventilation performance.

Keywords: CFD, HVAC, education.

1 Introduction

Computational Fluid Dynamics (CFD) which utilizes numerical methods to solve differential equation of describing fluid motion with the computer reveals the physical law of the flow. A key advantage of CFD is that it is a very compelling, non-intrusive, virtual modeling technique with powerful visualization capabilities, and engineers can evaluate the performance of a wide range of HVAC system configurations on the computer without the time, expense, and disruption required to make actual changes onsite.

So far, CFD is mainly used for the following five domains in the area of Heating, Ventilating and Air Conditioning (HVAC):

(1) The numerical simulation of the natural ventilation: American MIT, Hong Kong University study natural ventilation by virtual of Large Eddy Simulation (LES).

(2) The numerical simulation of the displacement ventilation: the displacement ventilation on the floor was studied by American MIT, S.V.Patankar [1] in Aalborg University in Denmark, Li Qiangmin [2] and Zhao Zhichao [3] in Tsinghua University in China.

---

[*] Corresponding author.

K. Li et al. (Eds.): LSMS / ICSEE 2010, Part I, CCIS 97, pp. 238–244, 2010.

(3) The numerical simulation of the huge space: the space air diffusion and the air conditioning load of the huge space such as gymnasium was studied by Zhao Deqing in Tsinghua University, Fan Cunyang [4], Ooi Yongson [5], Liu Fang [6] and Dong Yuping [7]. Hai Ying investigated the flow field between high constructions. Li Huizhi simulated the flow field of the wind tunnel between buildings in city.

(4) The numerical simulation of the Volatile Organic Compounds (VOC). With CFD, American MIT studied the diffusion of VOC and Indoor Air Quality (IAQ) in the house.

(5) The numerical simulation of the Clean Room: Tsinghua University simulated the air distribution of the air conditioning of the Clean Room to conduct the engineering design.

Base on the advantage of CFD, it is necessary to introduce it to HVAC education in college. By solving fundamental equations, CFD modeling gives students a detailed description of the fluid flow, heat transfer, and chemical species transport in the HVAC system or component. The students gain the understanding that they need for efficient troubleshooting, and insight that is difficult or impossible to get from experimental programs or field tests. It would be of great significance for theHVAC professional undergraduate students to master the CFD method. Then, one practical example was shown how the students to learn more from the CFD simulation work.

2 The Models Created by the Students

In order to help the student understand easily, we chose one classroom as the model created in CFD software. Most school buildings in China are designed for natural ventilation. It always leads to poor thermal environment in summer. Therefore, it is important to find an effective method to improve the thermal environment in school buildings. Considering the cost, installing fans is a useful method to improve ventilation performance in school buildings in China. Since the size of classrooms is always large, the installing position and quantity of the fans can affect the ventilation performance. Accurate CFD simulations of the flow field in the building can easily and quickly provide economical predictions of the poor zones. Therefore, we require the students to find effective ways to improve ventilation performance in school buildings by fans.

Generally, the plan design of classrooms in the school building in China is identical. So, a typical classroom located in Tianjin Polytechnic University was chosen to simulate. The CFD software we used in this study is FLUENT. In this study, three cases, which are fans without working, two fans working and four fans working, are respectively provided for the simulation in which boundary conditions used are measured on-site.

2.1 The Description of Model Dimensions

The selected classroom with length 15.75m along east-west direction, width 15.25m along south-north direction, height 3.5m, has 6 windows in the south wall, 5 windows in the north wall. The dimension of windows is 160×140(cm). And there are two doors in the north wall and the east wall respectively, whose dimension of is 220×140(cm). While ventilating, the doors are full open and the windows are half open. 8 rows and 3 columns tables are located in the classroom, in which the dimension of tables located opposite sides of the classroom is 250×37(cm), and the dimension of middle tables is 500×37(cm). The height of all tables is 80(cm). Table 2 presents all the dimensions of the CFD model. The radius of the ceiling fans is 50cm. The velocity of the flow through the windows is 0.3 m/s.

Figure 1 shows the model created which included the structure of the classroom, the location of the ceiling fans and the grids. As the inlets and outlets, the dimension of windows in the model corresponds to half of the real ones for that they are sliding windows and the dimension of doors correspond to the real ones.

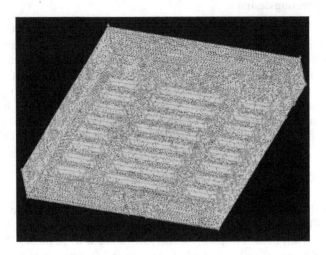

Fig. 1. CFD model created in this study

2.2 CFD Simulation Modeling of Turbulence

The governing equations for turbulent reacting flows are the Navier–Stokes Equations (NSE). The physical parameters such as air temperature, air pressure and air velocity in the turbulence randomly vary with the time and space. In physical structure, the turbulence is regarded as the flow consisted of diverse eddy of which the size and the axis of rotation are random.

One-equation model

$$\rho\frac{\partial k}{\partial t} + \rho u_j \frac{\partial k}{\partial x_j} = \frac{\partial}{\partial x_j}\left[\left(\eta + \frac{\eta_t}{\sigma_k}\right)\frac{\partial k}{\partial x_j}\right] + \eta_t \frac{\partial u_j}{\partial x_i}\left(\frac{\partial u_j}{\partial x_i} + \frac{\partial u_i}{\partial x_j}\right) - c_D\rho\frac{k^{3/2}}{l} \tag{1}$$

Where σ_k is the Prandtl (≈ 1.0) of the fluctuation kinetic energy, coefficient c_D in the references does not agree with each other. However, when determine η_t with $k - \varepsilon$ model, we only care for the product of c_D and c_μ'. And the product in various references is considerable identical (≈ 0.09).

$k - \varepsilon$ two-equation model

The equation of dissipation rate ε is:

$$\rho\frac{\partial\varepsilon}{\partial t}+\rho u_k\frac{\partial\varepsilon}{\partial x_k}=\frac{\partial}{\partial x_k}\left[\left(\eta+\frac{\eta_t}{\sigma_\varepsilon}\right)\frac{\partial\varepsilon}{\partial x_k}\right]+\frac{c_1\varepsilon}{k}\eta_t\frac{\partial u_i}{\partial x_j}\left(\frac{\partial u_i}{\partial x_j}+\frac{\partial u_j}{\partial x_i}\right)-c_2\rho\frac{\varepsilon^2}{k} \qquad (2)$$

So the k equation could be rewritten:

$$\rho\frac{\partial k}{\partial t}+\rho u_j\frac{\partial k}{\partial x_j}=\frac{\partial}{\partial x_j}\left[\left(\eta+\frac{\eta_t}{\sigma_k}\right)\frac{\partial k}{\partial x_j}\right]+\eta_t\frac{\partial u_i}{\partial x_j}\left(\frac{\partial u_i}{\partial x_j}+\frac{\partial u_j}{\partial x_i}\right)-\rho\varepsilon \qquad (3)$$

Where c_1, c_2 are empirical coefficients. The recommended value of the coefficients in $k - \varepsilon$ model is given in Table 1.

Table 1. The coefficients in $k - \varepsilon$ model

c_μ	c_1	c_2	σ_k	σ_ε	σ_T
0.09	1.44	1.92	1.0	1.3	0.9~1.0

3 Results and Discussion

In order to determine the ventilation performance in the classroom, two surfaces were chosen for the comparison of the simulation results.Surface-1, the horizontal plane which is 1m vertical distance from the floor has been chosen to examine the air velocity performance because this height is students' sitting height. Surface-2, the vertical face directing south and north under the ceiling fans has been chosen to compare the air velocity on high level in the classroom.

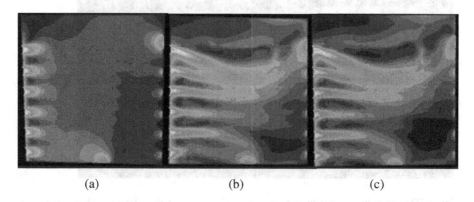

(a)   (b)   (c)

Fig. 2. The air velocity distribution on surface-1: (a) the ceiling fan is not open; (b) two ceiling fans are open; (c) four ceiling fans are open

Figure 2 (a) shows the air velocity distribution without fans working. It can be seen that when the ceiling fans are closed, the ventilation of the row adjacent to south windows (inlets) is the best while the ventilation in the northeast of the classroom is the worst. The area of the east and southeast of the classroom is between them. The middle and northwest is worse than the east and southeast. The southwest and northeast of the classroom exists large dead zone. Generally, the ventilation of the area through southeast and northwest is relatively better.

Figure 2 (b) shows the air velocity distribution with two ceiling fans working. When the two ceiling fans in front of the classroom are open, the ventilation from the first south window to the north door is good while air velocity decreases by degrees eastwardly and increases by degrees at the southeast. It quickly damps westward. The four corners in the classroom all appear dead zone. The dead zone at the platform becomes larger while the ventilation at the north platform gets better.

Figure 2 (c) shows the air velocity distribution with four ceiling fans working. When the four ceiling fans are all open, the dead zone at the northeast corner and southwest corner become larger while the ventilation at southeast get better than the case that only two ceiling fans are open.

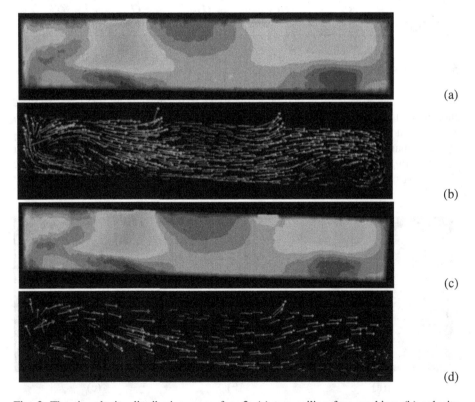

(a)

(b)

(c)

(d)

Fig. 3. The air velocity distribution on surface-2: (a) two ceiling fans working; (b) velocity vectors colored by velocity magnitude when two ceiling fans are open; (c) four ceiling fans working; (d) velocity vectors colored by velocity magnitude when four ceiling fans are open.

Figure 3 presents the air velocity distribution on surface 2. From the figure, it can be found that dead zone exists between the ceiling fans and it enlarges when four fans are open. Moreover, the dead zone is larger four fans opening than two fans opening. Large flow distribution exists in the area close to the south window and below the two fans. Although the dead zone exists below the desk, it does not matter due to the fact that it prevents students' legs from direct blow by the fans. The dead zone and eddy could also be obviously observed in Figure 3(b) and (d).

Because the dead zone between fans belongs to the area of assembly occupancy, the ventilation should be improved in this area. The northeast dead zone is the corner of the classroom and it has little effect on students. So it is not necessary to tackle the northeast dead zone. One measure installing one more fan in the middle of the classroom is selected after many simulations. Figure 4 presents the effect of the ventilation after being improved. Obviously, the dead zone between fans vanishes, which satisfies our prediction. This is one of the improved solutions. Besides, we could install exhaust fans in north wall, or reinstall all the ceiling fans, etc.

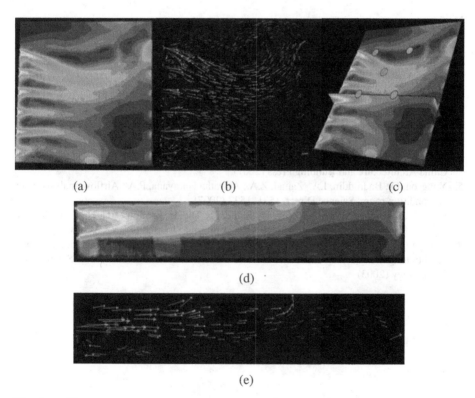

Fig. 4. (a) The result of the simulation on surface-1 after being improved; (b) velocity vectors colored by velocity magnitude on surface-1 after being improved; (c) the result of the simulation on surface-2; (d) velocity vectors colored by velocity magnitude on surface-2; (e) the assembly drawing

4 Conclusion

Through the application of CFD software, the students got more information on the CFD method. In the simulation, the students observe that the ventilation with four fans working is better than the case when two fans are open. However, the dead zone in the middle of the classroom becomes larger when four fans are open. And this area is just the assembly area. So it react the unreasonable fans installing. One improvement is provided in this study. After improving, the classroom has five fans, and the dead zone in the middle of the fans vanishes.

Moreover, the students can draw a conclusion that the simulation result and the research method could provide basis for improvement and reference for installing fans in new school buildings and the similar architectures designed for natural ventilation. The method integrating simulation and investigation could be used in the similar study. It is very useful for HVAC students to conduct this software in their course study.

References

1. Patanka, S.V.: Heat transfer and numerical computation of flow, Hemisphere, Washington, DC (1980)
2. Li, Q., Liangping, R.: The analysis based on the design philosophy and operating conditions of the energy saving in displacement ventilation. Refrigeration Air Conditioning & Electric Power Machinery (2004)
3. Zhichao, Z., Ziqiang, S.: The numerical simulation and analysis of displacement ventilation. Refrigeration & Air-Conditioning 4 (2004)
4. Cunyang, F.: The air-conditioning design and documentation of the large space building. China Architecture and Building Press (2001)
5. Yongson, O., Badruddin, I.A., Zainal, Z.A., Aswatha Narayana, P.A.: Airflow analysis in an air condition room. Science Direct, 1531–1537 (2007)
6. Fang, L., Miaocheng, W., Qingping, P., Junjie, Z.: The application of CFD in the air-flow organization design of departure lonnge air-conditioning. Journal of HV&AC (2007)
7. Yuping, D., Shijun, Y., Hongjun, W., Ziping, Z., Xiaofen, R.: The CFD simulation of air-conditioning flow in large space. Journal of Hebei Institute of Architectural Science and Technology (2003)

Power-Aware Replacement Algorithm to Deliver Dynamic Mobile Contents

Zhou Su[1] and Zhihua Zhang[2]

[1] Faculty of Science and Engineering, Waseda University,
Ohkubo3-4-1, Tokyo, Japan
`zhousu@asagi.waseda.jp`
[2] Department of Human Life Studies Sanyo Women' College,
Hiroshima 738-8504, Japan
`zhang@sanyo.ac.jp`

Abstract. As more and more users are using wireless network to access Web contents, the power awareness issue becomes one of the most important concerns for Mobile contents delivery networks (MCDN). Unnecessary power dissipation always brings the disconnection and delay during the time of wireless access. Replacement algorithm is looked upon as one solution to resolve this problem. However, most of the current replacement algorithms have not been taken the power-awareness into consideration. Therefore, in this paper we design a new method where a novel power-aware algorithm is proposed. Both theory analysis and simulations improve that our proposal can outperform other conventional methods.

Keywords: Mobile Contents, Contents Processing, Network Architecture, Power Aware, Replacement Algorithm.

1 Introduction

Mobile contents delivery network (MCDN) has emerged as a promising alternation to the conventional client-server based networks, by distributing the replicas of mobile contents onto a group of mobile nodes geographically. However, with the recent progress of mobile technology, more and more dynamic contents are being delivered by mobile users and need to be controlled [3]-[10].

For example, with posting the dynamically changing contents such as online auction or advertisements, because these contents are updated frequently on their original servers, how to replace the old version of these replicas on different mobile nodes becomes very important. If the unused contents can not be replaced with other new contents on time, it will cause a waste of power capacity since these contents may not be requested by users.

The above problem is called mobile contents replacement and plays an important role in the performance of the MCND. But the conventional replacement methods are almost for the wired network and can not be directly applied into the mobile environment.

K. Li et al. (Eds.): LSMS / ICSEE 2010, Part I, CCIS 97, pp. 245–250, 2010.

Therefore, this paper proposes a novel replacement algorithm for the MCDN. Note that this proposal is not designed for the consistency control to update the contents, while this paper is to decide which replica in a given mobile node should be removed completely from the mobile node to make room for the newly coming contents. We firstly make a theoretical analysis of the distribution of mobile contents. Then, based on the result of analysis, we present the proposed algorithm for replacement. Finally, we test the proposal by simulation experiments. The results show that our proposal can outperform other conventional method, where the capacity resource can be used more efficiently.

2 Theory Analysis

2.1 Parameters Definition

For each mobile node i ($i=1,...,$ I) in a MCDN, O_i is defined as its capacity and λ_i (bytes/second) denotes an aggregate request rate from the clients to this node. The total number of contents delivered in the MCDN is J. For each content j ($j=1,...,$ J), let h_j define the request probability that this content is requested by clients. Its data size is defined as b_j.

The Request Routing (RR) function is available in each node in the mobile networks, where this RR function keeps the residence time value $x_{q,i}$ of each client q ($q =1,...,$ Q) in node i 's zone. The local data provided by the RR function can be described by

$$X = \begin{bmatrix} x_{1,1} & \cdots & x_{1,I} \\ \cdots & x_{q,i} & \cdots \\ x_{Q,1} & \cdots & x_{Q,I} \end{bmatrix} \tag{1}$$

A matrix G represents the placement pattern of different contents on different nodes by:

$$G = \begin{bmatrix} g_{1,1} & \cdots & g_{1,J} \\ \cdots & g_{i,j} & \cdots \\ g_{I,1} & \cdots & g_{I,J} \end{bmatrix} \tag{2}$$

where $g_{i,j}$ in the above matrix takes a binary value as follows.

$g_{i,j} = 1$ (if the replica of content j is available in node i)

$$g_{i,j} = 0 \text{ (otherwise)} \tag{3}$$

Assume that the exhausted power to fetch the content j from its original sever is $C_{i,j,q}$, in the case that content j is not available in the node i, which it is requested by client q. Then we can get the total exhausted power to be:

$$\sum_{j=1}^{J} (g_{i,j} \cdot C_{i,j,q}) \qquad (4)$$

2.2 Analysis of Contents Delivery

Studies [1] show that the distribution of Web accesses follows a Zipf distribution, where the probability that the content j is requested can be obtained as follows:

$$h_j = \frac{\sum_{j=1}^{J} \dfrac{1}{w_j^{\theta}}}{w_j^{\theta}} \qquad (5)$$

Here, θ is a parameter of the Zipf distribution, and w_j denotes the ranking based on the requesting frequency.

According to the RR function, as for each node i in the MCDN, the relative residence of clients within node i can be obtained by

$$x_i = \frac{\sum_{q=1}^{Q} x_{q,i}}{\sum_{q=1}^{Q} \sum_{i=1}^{I} x_{q,i}} \qquad (6)$$

If a client firstly enters in the area of node i and then continue to request contents j, we can get the probability that the content j is requested from node i to be:

$$\alpha_{i,j} = \frac{\sum_{j=1}^{J} \dfrac{1}{w_j^{\theta}} \cdot \sum_{q=1}^{Q} x_{q,i}}{w_j^{\theta} \cdot \sum_{q=1}^{Q} \sum_{i=1}^{I} x_{q,i}} \qquad (7)$$

2.3 Proposed Algorithm

In our method, if the node's capacity exceeds its limit and the node decides to replace some content with the newly coming one, the content i will be selected according to:

$$\text{If} \begin{cases} \alpha_{i,j} = \min(\alpha_{i,j}) \\ \sum_{j=1}^{J} g_{i,j} \cdot b_j > O_i \end{cases} \qquad (8)$$

Then $g_{i,j} = 0$

That is to say: when a node i exceeds its capacity limitation, the content j in node i which has the lowest value based on Eq.6 will be selected to be removed from the node to save capacity for the newly coming contents.

Based on the replacement carried by Eq.8, we can obtain the new placement of contents by

$$G' = \begin{bmatrix} g'_{1,1} & \cdots & g'_{1,J} \\ \cdots & g'_{i,j} & \cdots \\ g'_{I,1} & \cdots & g'_{I,J} \end{bmatrix} \tag{9}$$

Since we have denoted $C_{i,j,q}$ as the exhausted power to fetch the content j from its original sever, we can obtain the total exhausted power when the placement of contents is changed by the proposed replacement algorithm as follows.

$$\sum_{j=1}^{J} (g'_{i,j} \cdot C_{i,j,q}) \tag{10}$$

Finally the reduced amount after using the proposal becomes:

$$\sum_{j=1}^{J} (g'_{i,j} \cdot C_{i,j,q}) - \sum_{j=1}^{J} (g_{i,j} \cdot C_{i,j,q}) \tag{11}$$

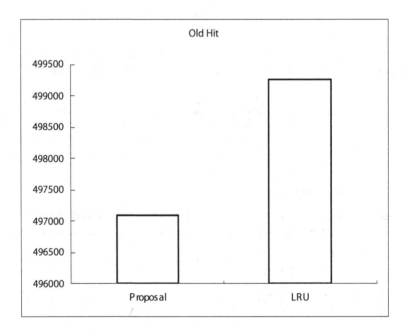

Fig. 1. Result of Old hit

3 Simulation

We evaluate our proposal method by the following simulation condition. All of the nodes in the MDCN are placed according to the Power-Law distribution [2]. The request arrives according to a Poisson process and the popularity of each content is decided by the Zipf distribution [1]. The update period of each content is decided at random. The capacity of each node is set to be 2% of the total size of all contents delivered in the MCDN. The simulation times is 1000000. We compare our proposal with the conventional *LRU* method.

Figure 1 shows the result of *Old Hit* when different algorithms are carried out. Here, the *Old Hit* means the situation that the node keeps an old version of the user's requested content. From the result in Fig.1, we can know that our proposal can obtain a lower *Old Hit* than the *LRU*. As a result, the user can be satisfied with taking the contents from the node directly, instead of contacting the remote original server to fetch the contents, resulting in a low user delay. We continue to test the performance of network traffic. It is because that the requested content can be provided to the user from the nearby MCDN nodes, then the traffic which is caused by fetching data from original servers to the users is reduced.

4 Conclusion

In this paper, we proposed a power-aware replacement algorithm for mobile contents delivery. Results showed that both better hit ratio and network traffic can be obtained compared with the conventional method. More theoretical analysis and implementation will be carried out as the future work.

References

1. Breslau, L., Cao, P., Fan, L., Phillips, G., Shenker, S.: Web Caching and Zip-like Distributions: Evidence and Implications. In: IEEE INFOCOM 1999, New York (1999)
2. Stoica, I., Morris, R., Karger, D., Kaashoek, M.F., Chord, H.B.: A Scalable Peer-to-peer Lookup Service for Internet Applications. In: ACM SIGCOMM 2001, pp. 27–31 (2001)
3. Zhou, S., Katto, J., Yasuda, Y.: Scalable Maintenance for Strong Web Consistency in Dynamic Content Delivery Overlays. In: Proceedings of 2007 IEEE ICC 2007, Scotland (2007)
4. Li, X., Ji, H., Zheng, R., Li, Y., Yu, F.R.: A Novel Team-centric Peer Selection Scheme for Distributed Wireless P2P Networks. In: IEEE WCNC 2009 Budapest, Hungary (2009)
5. Ding, J.W., Lan, S.Y.: Quality-Adaptive Proxy Caching for Peer-to-Peer Video Streaming Using Multiple Description Coding. Journal of Information Science and Engineering 25(3), 687–701 (2009)
6. Ding, J.W., Tseng, S.Y., Huang, Y.M.: Packet Permutation: A Robust Transmission Scheme for Continuous Media Streams over the Internet. Multimedia Tools and Applications 21(3), 281–305 (2003)
7. Su, Z., Oguro, M., Katto, J., Yasuda, Y.: Selective Update Approach to Maintain Strong Web Consistency in Dynamic Content Delivery. IEICE Trans. on Commun. (October 2007)

8. Ortega, A., Carignano, F., Ayer, S., Vetterli, M.: Soft Caching: Web cache management techniques for images. In: IEEE Signal Proc. Society Workshop on Multimedia Signal Processing, Princeton, NJ (June 1997)
9. Arlitt, M., Friedrich, R., Jin, T.: Performance evaluation of web proxy cache replacement policies. In: Puigjaner, R., Savino, N.N., Serra, B. (eds.) TOOLS 1998. LNCS, vol. 1469, p. 193. Springer, Heidelberg (1998)
10. Shen, H., Joseph, M.S., Kumar, M., Das, S.K.: PReCinCt: A Scheme for Cooperative Caching in Mobile Peer-to-Peer Systems. In: Proceedings of the 19th IEEE International Parallel and Distributed Processing Symposium (2005)

Study on High-Frequency Digitally Controlled Boost Converter

Yanxia Gao[1], Yanping Xu[1], Shuibao Guo[1,2], Xuefang Lin-Shi[2], and Bruno Allard[2]

[1] School of Mechatronical Engineering and Automation, Shanghai University
200072 Shanghai, China
{gaoyanxia,xuyanping}@shu.edu.cn, shuibaoguo@gmail.com
[2] Lab. AMPERE (CNRS UMR 5005)-INSA-Lyon, Villeurbanne Cedex – France
{xuefang.shi,bruno.allard}@insa-lyon.fr

Abstract. This paper presents a completely digitally controlled high-frequency boost converter. The research focuses on the two key modules: Digital Pulse-Width Modulation (DPWM) and digital control-law. The proposed hybrid DPWM architecture, which takes advantage of Digital Clock Manager (DCM) phase-shift characteristics available in FPGA resource and combines a counter-comparator with a digital dither block, is introduced firstly, and then a digital control algorithm is designed. Finally, based on a Xilinx Virtex-II Pro FPGA board with 32MHz hardware clock, an 11-bit DPWM and a digital controller are implemented for a boost converter. The performance of the converter is validated by experimental results.

Keywords: Boost Converter, Digital Control, Hybrid DPWM, FPGA Implementation.

1 Introduction

With the development of low-power portable electronics devices and embedded systems, Switching Mode Power Supplies (SMPS) are demanded to meet the critical requirements, such as higher performance, smaller size and higher efficiency. Especially, the miniaturization becomes a design issue for integrated SMPS. In order to reduce the size of SMPS, high switching frequency operation is essential to reduce size of passive components. Thus, it is difficult for traditional analogy controlled SMPS to meet these critical requirements. By contrast high-frequency digital control technology for SMPS application has become an attractive research in recent years. Compared with analog control, digital control technique has advantages of advanced control algorithms application, flexibility, programmability, less sensitivity to variations and has a trend of cost reduction [1-3]. Fig. 1 shows a diagram block of digitally controlled boost converter.

DPWM module and digital compensator design are the two main issues for SMPS digital control realization. The purpose of the research is to achieve high resolution with high switching frequency and low system clock frequency. The general DPWM methods are dither, delay-line and sigma-delta [4-6]. There are merits and drawbacks

K. Li et al. (Eds.): LSMS / ICSEE 2010, Part I, CCIS 97, pp. 251–258, 2010.

for each method respectively [7]. In this paper, a novel hybrid DPWM is presented to acquire high-frequency high-resolution PWM under a low-frequency hardware clock so as to solve the constraint between DPWM resolution and system frequency on one hand, and to guarantee precise output voltage and low power consumption on other hand. For the control law, advanced strategies can be applied to achieve better performance in digital control [8], while the procedure to obtain the controller parameters relies on a complex optimization algorithm which will cause high cost. This paper presents the design and application of a digital control algorithm by using Matlab. For the validation, the proposed digital controller is implemented on a boost converter with a Xilinx FPGA, and experimental results verify the proposal.

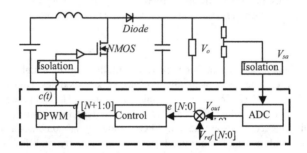

Fig. 1. Structure of digitally controlled boost SMPS system

2   Proposed 11-Bit FPGA-Based Hybrid DPWM

The proposed DPWM includes three blocks: 3-bit digital dither block, 4-bit segmented DCM phase-shift block and 4-bit counter-comparator block. Fig. 2 shows the schematic blocks of the proposed DPWM architecture.

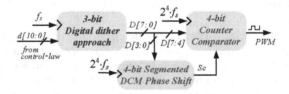

Fig. 2. Proposed DPWM schematic block

2.1   3-Bit Digital Dither Block

The basic principle of digital dither is detailed in [4]. It consists to distribute the Ndith LSB of the duty ratio in a pre-scheduled sequence and put the specific LSB effects into hardware NDPWM MSB. The (NDPWM + Ndith) bits duty ratio from control law will be modified in an average distribution over 2Ndith switching periods, so that the equivalent duty ratio is in the value between 2Ndith adjacent quantized levels. By digital dither method, the Core DPWM resolution NDPWM can be increased by Ndith bits up to equivalent NDPWM +Ndith bits.

However, dither is not coming free. The longer bits the dither is used, the higher output ripple increases. Thus, this consideration puts a practical limit on the number of dither bits that can be added to increase the resolution of the DPWM. When the digital dither approach is applied to the proposed 11-bit DPWM architecture, the bit number of dither can be determined using those useful mathematical analyses in [4]. According to the determination, a 3-bit digital minimum-ripple dither pattern is adopted in the proposed 11-bit hybrid DPWM.

Fig. 3 and Fig. 4 show the diagram block of 3-bit digital dither and its minimum-ripple dither scheme, respectively. As shown in Fig. 4, where d1 and d2 are two adjacent initial quantized levels with d1 = d2 + LSB. It can be seen that when the duty ratio changes between d1 and d2 in a dither sequence during every 23 switching periods, a corresponding sub-bit level can be implemented by averaging over 8 switching periods. According to Fig. 4, a look-up table should be used to store the 23 dither sequences. Each sequence is 23-bit long. The dither value will be added to the d[10:3] by an 8-bit saturated adder, which generates a new duty ratio D[7:0]. As a result, the equivalent resolution of DPWM is increased by 3 bits.

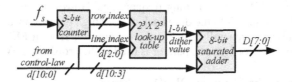

Fig. 3. A diagram block of 3-bit digital dither

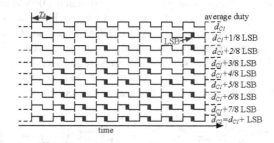

Fig. 4. A 3-bit digital dither minimum-ripple dither scheme

2.2 4-Bit Segmented DCM Phase-Shift Block

DCM is available in most FPGA devices. It can implement a clock delay locked loop, a digital frequency synthesizer and digital phase shifter. Here, DCM shifts the clock phase optionally to delay the incoming clock by a fraction of the clock period. For instance shown in Fig. 5, the DCM divides the incoming clock FCLK (50% ratio) into four equal clocks clk_0, clk_90, clk_180 and clk_270 respectively, then the four phase-shifted clocks can act as an equivalent 22•FCLK clock with a 4:1 multiplexer.

Thus, the clock for the DCM architecture can be reduced by 22 times for a fixed-resolution, or the resolution can be increased by 2 bits for a fixed frequency.

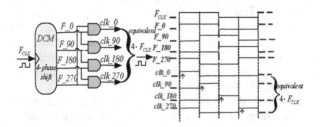

Fig. 5. DCM four-phase-shift scheme

2.3 4-Bit Counter-Comparator Block

Counter-comparator is one traditional approach to achieve digital-to-time conversion in DPWM application. This scheme has the advantage of a simple structure and an excellent linearity in the digital-to-time conversion. According to [5], it needs 2N•fs clock to achieve an N-bit DPWM at switching frequency fs. However, when it operates at the high frequency fs, it falls into the drawback of very high power consumption. Thus, the counter-comparator is generally used as the solution for few-bit MSBs inside DPWM architectures. A 4-bit counter-comparator is used in the proposed DPWM.

2.4 Operation of the Proposed DPWM Scheme

Taking a combination of three blocks described above: digital dither block, DCM phase-shift block and counter comparator, the completed DPWM architecture can be figured in Fig.6. Among the 11-bit DPWM, 3-bit are implemented as delta-sigma modulator, 4-bit are achieved by segmented DCM phase-shift block, and 2-bit are generated by counter-comparator. According to the proposed DPWM architecture, when an 11-bit DPWM operates at fs switching frequency, the system merely needs 24•fs hardware clock instead of 211•fs. Clearly the proposed hybrid DPWM dramatically alleviates the clock requirement, and can reduce power consumption.

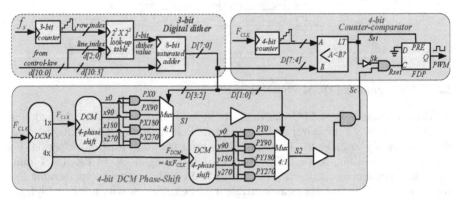

Fig. 6. Proposed 11-bit FPGA-based DPWM

3   The Design of Control Law

The transfer function of the Boost converter can be gained by averaging state-space method and small-signal analysis in continuous current mode (CCM), which can be described as below:

$$G_{vd}(s) = \frac{\hat{v}_o(s)}{\hat{d}(s)}\bigg|_{\hat{v}_i(s)=0} = \frac{(1+sR_cC)[R(1-D)V_o - (R_E+sL)V_o / (1-D)]}{R_E + R(1-D)^2 + s[L + R_E(R_c+R)C + R_cR(1-D)^2C] + s^2(R+R_c)LC} \tag{1}$$

Where, $R_E = R_L + DR_{on} + (1-D)R_F$ R_L, R_C is the equivalent series resistance of inductor and capacitor, respectively. R_{on}, R_F is the on-resistance of transistor and diode, respectively. IL is averaged inductor current, and D is PWM duty ratio.

Substituting the circuit parameters into equation (1), with $V_i = 3V$, $V_o = 6V$, $R = 50\Omega$, $L = 100\mu H$, $C = 100\mu F$, $V_F = 0.44V$, $R_F = 0.027\Omega$, $R_L = 0.22\Omega$, $R_c = 0.2\Omega$, $R_{on} = 0.0037\Omega$, then the transfer function of actual boost converter $G_{vd}(s)$ is obtained:

$$G_{vd}(s) = \frac{-1.2\times10^{-8}s^2 + 6.21\times10^{-4}s + 61.05}{2.29\times10^{-7}s^2 + 6.773\times10^{-4}s + 4.856} = \frac{-0.052391(s-1.017\times10^5)(s+5\times10^4)}{s^2+2957s+2.12\times10^7} \tag{2}$$

With Sisotool in Matlab, pole-zero placements is realized and the transfer function of a comparatively ideal compensator is acquired:

$$G_c(s) = \frac{1.62\times10^6(s+4.62\times10^3)(s+6.21\times10^3)}{s(s+9.94\times10^4)(s+3.1\times10^5)} \tag{3}$$

Transfer equation (3) into discrete-time z-domain and the digital PID controller can be expressed as,

$$G_c(z) = \frac{1.725z^3 - 1.651z^2 - 1.724z + 1.652}{z^3 - 1.918z^2 + 1.085z - 0.1671} \tag{4}$$

The digital controller can be described as:

$$d[k] = 1.725e[k] - 1.651e[k-1] - 1.724e[k-2] + 1.652e[k-3]$$
$$+ 1.918d[k-1] - 1.085d[k-2] + 0.1671d[k-3] \tag{5}$$

Where $d[k]$ is the discrete value of the output of PID controller, $e[k]$ is the discrete value of the error between reference voltage and sampling voltage, $d[k-i]$ and $e[k-i]$ are the values in the ith-cycles before the current cycle respectively. Thus, the control arithmetic can be realized by VHDL on FPGA.

4   The Experimental Results

The functionality of the controller designed and proposed DPWM are experimentally verified using a discrete boost converter with 3.0V input and 6.0V output voltage. An 11-bit discrete ADC (AD9237) is used in order to avoid limit cycle oscillation. The implementation of the proposed digital controller DPWM-based is performed on a

Xilinx XC2VP30 FPGA with an external 32MHz system clock. A VHDL design approach is used to synthesize the controller with Xilinx ISE development software. The test platform is pictured in Fig.7.

Fig. 7. Experimental test platform

Fig.8 (a) shows the steady-state output voltage and the corresponding PWM signal. It can be seen that the system has favorable steady-state performance and the output ripple voltage is less than 60mV. Fig.8 (b) shows the steady-state inductor current.

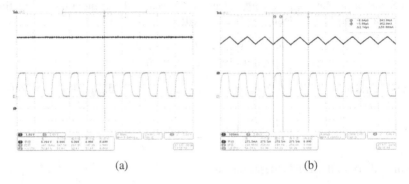

(a)   (b)

Fig. 8. Steady-state waveform (a) Output voltage (b) Inductor current

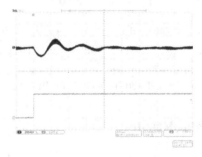

Fig. 9. Transient response waveform of output voltage and the corresponding load

Fig.9 shows the transient response waveform of output voltage and current when load varies from 50Ω to 25Ω (current from 0.12A to 0.24A). It can be seen that the overshoot is about 160mV and the regulation time is 1.6ms. It shows the satisfied dynamic response performance of digital controller and verifies the correctness of simulation design in part3.

5 Conclusion

This paper presents a digital controller for high-frequency low-power DC-DC SMPS. The architecture contains a Hybrid DPWM and a low-cost PID algorithm module. The Hybrid DPWM architecture takes advantage of DCM phase-shift characteristics available in FPGA resource and combines a counter-comparator block with digital dither modulator. The proposed DPWM greatly alleviates the constraint between PWM resolution and system clock frequency. Along with a PID algorithm, the proposed digital controller is implemented by Xilinx XC2VP30 FPGA board on a discrete synchronous low-power boost converter at switching frequency of 250KHz. The experimental results verify the design.

Acknowledgments

This research is co-supported by Shanghai University "11th Five-Year Plan" 211 Construction Project, Shanghai Key Laboratory of Power Station Automation Technology and Power Electronics Science Education Development Program of Delta Environmental & Education Foundation (No.DERO2007014).

References

1. Patella, B., Prodić, A., Zirger, A., Maksimović, D.: High-frequency Digital PWM Controller IC for DC-DC Converters. IEEE Transactions on Power Electronics 18(1), 438–446 (2003)
2. Liu, Y.F., Sen, P.C.: Digital Control of Switching Power Converters. In: Proceedings of the 2005 IEEE Conference on Control Applications, pp. 635–640. IEEE Press, Toronto (2005)
3. Peng, H., Prodic, A., Alarcon, E., Maksimovic, D.: Modeling of Quantization Effects in Digitally Controlled DC-DC Converters. In: Proc. IEEE Power Electronics Specialists Conf., pp. 4312–4318. IEEE Press, Aachen (2004)
4. Peterchev, V., Sanders, S.R.: Quantization Resolution and Limit Cycling in Digitally Controlled PWM Converters. IEEE Transactions on Power Electronics 18(1), 301–308 (2003)
5. Huerta, S.C., de Castro, A., Garcia, O., Cobos, J.A.: FPGA Based Digital Pulse Width Modulator with Time Resolution under 2ns. In: Proc. of the IEEE APEC 2007, pp. 877–881. IEEE Press, Los Alamitos (2007)
6. Lukić, Z., Rahman, N., Prodić, A.: Multi-bit Sigma-delta Pwm Digital Controller ic for DC-DC Converters Operating at Switching Frequencies Beyond 10 mhz. IEEE Transactions on Power Electronics 22(5), 1693–1707 (2007)

7. Guo, S., Gao, Y., Xu, Y., Lin-Shi, X., Allard, B.: Digital PWM controller for high-frequency low-power DC-DC switching mode power supply. In: The 2009 IEEE 6th International Power Electronics and Motion Control Conference, pp. 1340–1346. IEEE Press, China (2009)
8. Lin-Shi, X., Morel, F., Allard, B., Tournier, D., Rétif, J.M., Guo, S., Gao, Y.: A Digital-Controller Parameter-Tuning Approach, Application to a Switch-Mode Power Supply. In: The 2007 IEEE International Symposium on Industrial Electronics, pp. 3356–3361. IEEE Press, Spain (2007)

Author Index